BestMasters

Springer awards „BestMasters" to the best master's theses which have been completed at renowned universities in Germany, Austria, and Switzerland.

The studies received highest marks and were recommended for publication by supervisors. They address current issues from various fields of research in natural sciences, psychology, technology, and economics.

The series addresses practitioners as well as scientists and, in particular, offers guidance for early stage researchers.

Verena Puchner

Evaluation of Statistical Matching and Selected SAE Methods

Using Micro Census and EU-SILC Data

Springer Spektrum

Verena Puchner
Vienna, Austria

BestMasters
ISBN 978-3-658-08223-9   ISBN 978-3-658-08224-6 (eBook)
DOI 10.1007/978-3-658-08224-6

Library of Congress Control Number: 2014956551

Springer Spektrum

Printed on acid-free paper

Springer Spektrum is a brand of Springer Fachmedien Wiesbaden
Springer Fachmedien Wiesbaden is part of Springer Science+Business Media
(www.springer.com)

Danksagung

Das Gefühl eine vollendete Diplomarbeit vor sich zu haben ist großartig, aber noch schöner ist es, zu wissen, dass ich so liebe, hilfsbereite, unterstützende und aufmunternde Menschen in meiner Nähe habe, die mich während des gesamten Verlaufes meines Studiums und meiner Diplomarbeit begleitet haben.

Bedanken möchte ich mich hiermit bei all meinen Verwandten, Freunden, Bekannten und auch Experten, die mir geholfen haben, nun an dieser Stelle angelangt zu sein.

Mein herzlicher Dank gilt Herrn Privatdoz. Dipl.-Ing. Dr.techn. Matthias Templ, dem Betreuer meiner Diplomarbeit, der den Prozess meiner Diplomarbeit von Anfang an sehr angenehm und bemüht leitete, mir auf jede noch so kleine Frage sehr rasch antwortete und mir stets hilfreich zur Seite stand. Vielen Dank auch für das geweckte Interesse in diversen Gebieten der Statistik.

An dieser Stelle möchte ich mich auch bei DI Alexander Kowarik für seine Hilfe und seinen kompetenten fachlichen Rat bedanken. Damit wurde mir das Erarbeiten und Verfassen meiner Diplomarbeit um ein Vielfaches erleichtert.

Ein großes Dankeschön ist an meine Eltern gerichtet, die mir dieses Studium überhaupt ermöglichten, immer ein offenes Ohr für mich hatten und mich jederzeit tatkräftig unterstützten. Bei meiner Mama möchte ich mich außerdem sehr herzlich für das gewissenhafte Korrekturlesen meiner Diplomarbeit bedanken.

Auf meinem Weg durch das Studium haben mich weiters meine drei Geschwister, Sandra, Julia und Tobias, begleitet, die stets Interesse an meiner Arbeit zeigten und mir die Zeit des Studierens durch Hilfe und Gesellschaft im Studentenheimleben sowie durch Tipps und Tricks die Lehrveranstaltungen betreffend erleichterten.

Meinen Studienkollegen, allen voran Irene, Barbara, Lisa, Alexander und Bernhard, möchte ich ebenfalls danken. Das Studieren miteinander, aber vor allem auch die Pausen zwischen der Arbeit und unsere gemeinsamen Aktivitäten, haben mich gestärkt, vorangebracht und diesen Lebensabschnitt für mich unvergesslich gemacht.

Ganz besonderer Dank gilt meinem Freund Michael, der mir starken Rückhalt gab, Verständnis zeigte, immer für mich da war und dessen liebevolle Unterstützung unschlagbar war.

Vielen herzlichen Dank!

Verena Puchner

Abstract

Due to the fact that poverty estimations on regional level on basis of EU-SILC samples is not of adequate accuracy, the quality of the estimations should be improved by additionally incorporating Micro Census data for estimation. In comparison to EU-SILC, the Micro Census survey data consists of more observations. However, income is not questionaired but necessary to estimate poverty.

The aim is to find the "best" method for the estimation of poverty in terms of small bias and small variance. Therefor an artificial "close-to-reality" population is simulated in order to know the "true" parameter values. To make an assessment of the quality, considering the respective sample designs, EU-SILC and Micro Census samples are drawn repeatedly. Variables of interest are imputed into the Micro Census data sets with the help of the EU-SILC samples through regression models including selected unit-level small area methods and statistical matching methods and poverty indicators are estimated in the following. The bias and variance for the direct estimator and the several methods are evaluated and compared. The variance is desired to be reduced by the larger sample size of the Micro Census. In conclusion, the result is that it doesn't exist only one method performing by far best in terms of bias and variance among the used models. Concerning the bias, most often the statistical matching methods perform better than the regression methods, but regarding the variances, the regression models do a better job in general. In terms of the average mean squared error of states the direct estimator performs best, followed by logistic (mixed) regression models and all the statistical matching methods.

Abstract

Contents

1. Introduction   1

2. Mathematical Background   3
  2.1. Definitions 3
  2.2. Distance Measures 4
  2.3. Regression Models Including Selected Small Area Methods 6
    2.3.1. Linear Regression 6
    2.3.2. Robust Linear Regression 9
    2.3.3. Logistic Regression 10
    2.3.4. Linear Mixed Regression 11
    2.3.5. Logistic Mixed Regression 11
    2.3.6. The Transmission of the Model 11
  2.4. Statistical Matching 12
  2.5. Bootstrap Methods 14

3. Design of the Simulation Study   17
  3.1. Problem 17
  3.2. Set-Up of the Design-Based Simulation Study 17
  3.3. Quality Criteria 20

4. Application to Poverty Estimation Using EU-SILC and Micro Census Data   23
  4.1. Data 23
    4.1.1. Details on Micro Census 23
    4.1.2. Details on EU-SILC 24
    4.1.3. Variables in the Population 24
  4.2. The Models 28
    4.2.1. Regression Models Including Selected Small Area Methods 28
    4.2.2. Statistical Matching 34
  4.3. Results 37
    4.3.1. Reliability of the Bootstrapping 37
    4.3.2. Bias 42
    4.3.3. Variance 47
    4.3.4. Mean Squared Error 51
    4.3.5. Bias Corrected Version 55

5. Conclusion   63

A. Appendix   65
  A.1. R-Code 65
    A.1.1. Regression Models Including Selected Small Area Methods 66
    A.1.2. Statistical Matching 78
    A.1.3. Additional R-Code 89
  A.2. Additional Tables 95

References   99

List of Figures

1.   The idea of constructing the artificial samples and evaluating the estimators. 18
2.   The sample design of Micro Census / data set A.   19
3.   The sample design of EU-SILC / data set B.   19
4.   Bar chart of the bias [in %] for the direct estimator (calculated on basis of
     the EU-SILC data sets)(denoted by S) and the several models (calculated
     on basis of the Micro Census data sets) using the example of Burgenland
     and Lower Austria.   46
5.   Bar chart of the standard deviation [in %] for the direct estimator (calcu-
     lated on basis of the EU-SILC data sets)(denoted by S) and the several
     models (calculated on basis of the Micro Census data sets) using the ex-
     ample of Burgenland and Lower Austria.   51
6.   Bar chart of the mean squared error for the direct estimator (calculated
     on basis of the EU-SILC data sets)(denoted by S) and the several models
     (calculated on basis of the Micro Census data sets) using the example of
     Burgenland and Lower Austria.   54

Listings

1. `sampling`: Sampling of **S** and **MZ** 65
2. `gewlmreg_soz`: Ordinary Least Squares Regression 66
3. `gewlmreg_rV`: Ordinary Least Squares Regression 67
4. `gewlmreg_sozrob`: Robust Linear Regression 68
5. `gewlmreg_rVrob`: Robust Linear Regression 69
6. `logreg_soz`: Logistic Regression 70
7. `logreg_rV`: Logistic Regression 71
8. `mixed_lm_inc_soz`: Linear Mixed Regression 73
9. `mixed_lm_inc_rV`: Linear Mixed Regression 74
10. `mixed_logm_arpt_soz`: Logistic Mixed Regression 76
11. `mixed_logm_arpt_rV`: Logistic Mixed Regression 76
12. `match_strat_arpt`: Random Hot Deck 78
13. `match_strat_inc`: Random Hot Deck 79
14. `match_stratmigr_arpt`: Random Hot Deck 80
15. `match_stratmigr_inc`: Random Hot Deck 81
16. `match_stratmigr_arpt_feklaw`: Sequential Random Hot Deck 82
17. `match_stratmigr_inc_feklaw`: Sequential Random Hot Deck 83
18. `match_stratmigr_arpt_ALLE`: Sequential Random Hot Deck 83
19. `match_stratmigr_inc_ALLE`: Sequential Random Hot Deck 84
20. `match_smr_strat_arpt`: Weighted Random Hot Deck: 86
21. `match_smr_strat_inc`: Weighted Random Hot Deck: 87
22. `match_smr_stratmig_arpr`: Weighted Random Hot Deck 87
23. `match_smr_stratmig_inc`: Weighted Random Hot Deck 88
24. `overall.R`: the R-Code for every node: 89
25. Parallelization and computation of quality criteria 93

List of Tables

1. Statistical Matching: the situation 12
2. Reliability of the bootstrapping regarding the variance on basis of the EU-SILC samples 37
3. Reliability of the bootstrapping regarding the variance on basis of the Micro Census samples 38
4. Reliability of the bootstrapping regarding the variance on basis of the Micro Census samples 39
5. Reliability of the bootstrapping regarding the at-risk-of-poverty rate estimations on basis of the EU-SILC samples 39
6. Reliability of the bootstrapping regarding the at-risk-of-poverty rate estimations on basis of the Micro Census samples 40
7. Reliability of the bootstrapping regarding the at-risk-of-poverty rate estimations on basis of the Micro Census samples 41
8. Models with the minimal / maximal absolute value of the difference in the means of the at-risk-of-poverty rates 42
9. At-risk-of-poverty rates of the artificial population [in %]. 42
10. Bias on basis of the EU-SILC samples [in %]. 43
11. Bias of the models for MZ 1 44
12. Bias of the models for MZ 2 45
13. Models with the smallest absolute value of the bias for Austria and the nine states. 47
14. Standard deviation on basis of the EU-SILC samples [in %]. 48
15. Standard deviation of the models for MZ 1 49
16. Standard deviation of the models for MZ 2 50
17. Models with the smallest variance for Austria and the nine states. 51
18. MSE on basis of the EU-SILC samples. 52
19. MSE of the models for MZ 1 52
20. MSE of the models for MZ 2 53
21. Models with the smallest MSE for Austria and the nine states 54
22. Corrected bias on basis of the EU-SILC samples [in %]. 55
23. Corrected bias of the models for MZ 1 56
24. Corrected bias of the models for MZ 2 57
25. Models with the smallest absolute value of the (corrected) bias for Austria and the nine states. 58
26. Bias corrected MSE on basis of the EU-SILC samples. 59
27. Bias corrected MSE of the models for MZ 1 60
28. Bias corrected MSE of the models for MZ 2 61
29. Models with the smallest (bias corrected) MSE for Austria and the nine states. 61
30. Average (of the nine states) of the difference in the mean of the standard deviations yield from the bootstrap replicates and the standard deviation calculated on the basis of the results of the repeated sample drawing (see also Equation 8 and Equation 9) for the several models [in %]. 95

31. Mean (of the several models) of the difference in the mean of the at-risk-of-poverty rates yield from the bootstrap replicates (of all drawn samples) and the mean of the at-risk-of-poverty rates estimated on the basis of the results of the repeated sample drawing (see also Equation 10 and Equation 11) for the nine states and Austria [in %]. 95

32. In the second column the average (of the nine states) of the bias for every model [in %] and in the third column the average (of the nine states) of the absolute values of the bias for every model [in %] are listed. In the fourth column the average (of the nine states) of the corrected bias for every model [in %] and in the fifth column the average (of the nine states) of the absolute values of the corrected bias for every model [in %] are listed. 96

33. Average (of the nine states) of the standard deviation for the direct estimator and every model [in %]. 97

34. In the second column the average (of the nine states) of the mean squared error for the direct estimator and every model and in the third column the average (of the nine states) of the bias corrected mean squared error for the direct estimator and every model are listed. 97

XIII

1. Introduction

This thesis deals with methods for the improvement of the quality of estimations of regional indicators using EU-SILC and Micro Census data sets.

The classical approach of estimating poverty is to estimate poverty indicators on basis of EU-SILC data sets. In the EU-SILC surveys information on income components are collected from which the equivalised household income is derived. In addition, the at-risk-of-poverty threshold and rate is estimated using the equivalised household income. Because of the relativ small sample size of EU-SILC, the variances of the estimates on regional level are large. Surveys from the Micro Census consists of more individuals and households, but income is not questionaired but necessary to estimate poverty.

However, if the equivalised household income can be transmitted on the Micro Census, the aim will be that the precision (in terms of mean squared or absolute errors) of estimating poverty on the basis of the EU-SILC data sets should be improved by using the larger sample size of the Micro Census. This is as much as to say, that thereby the uncertainty of the estimation respectively the variance should be reduced.

This is in particular important, in order to receive secured information at regional level. If this aim can be achieved, so among others the try to reduce the mean squared error (MSE) is successful, better results on regional level will be noticed and the quality of estimations on small domains will be higher.

Several methods of small area estimation, statistical matching and bootstrapping are used to impute the variables "equivalised household income" and "at-risk-of-poverty" into the (simulated) Micro Census data sets with the help of (simulated) EU-SILC-samples and to estimate poverty indicators in the following.

After diligent examination of the data, several models and methods, about 20 different small area models and statistical matching methods are applied to estimate the indicators of interest.

The intention is, to find the "best" method in matters of small bias and small variance.

First of all, an artificial "close-to-reality" population is simulated (containing both EU-SILC and Micro Census variables) to make it possible to compare the estimation results with the true parameter values.

In order to make an assessment of the bias and variance of the various estimation methods, considering the sample design of the surveys, Micro Census and EU-SILC samples are drawn repeatedly from this simulated population.

This approach is called design-based simulation, because from the simulated population samples are drawn considering the different sample designs. The alternative would be model-based simulation, where directly the samples, in this case the EU-SILC and Micro

Census samples, get simulated repeatedly. The advantage of the design-based simulation is that any design could be used, i.e. it is of interest to look how the estimators behave under different sampling designs.

The outline of the thesis is as follows: In the first chapter (Chapter 2) all the theoretical basics concerning the methods and models used are introduced. In the second chapter (Chapter 3) the problem, the set-up and the used quality criteria are outlined. In the third main chapter (Chapter 4) the methods are demonstrated on an concrete example. The underlying data set is explained, then the already explained methods and models are applied and the quality criteria are used for evaluation. In the end the results are summarized and in Chapter 5 prospects are given. The Appendix contains the R-Code used for the computations.

2. Mathematical Background

To explain some methods for combining two data sets consider the following situation: Let A and B be two sample surveys. The number of observations are different and not all variables in the two data sets are the same. Moreover some of the variables are observed in both surveys, some are observed in the sample survey A und some other variables are only available in the sample survey B. The idea is now to estimate the missing variables in one survey, lets say in A. Assume that this variable of interest has been observed in the survey B. Many different possibilities to perform this estimation exists. In this chapter the following kind of methods are considered:

1. Regression Models including selected unit-level Small Area Methods

2. Statistical Matching

For (1) linear regression, robust linear regression, logistic regression, linear mixed models and generalized linear mixed models (logistic mixed models) are evaluated.

For (2) random hot deck, sequential random hot deck and weighted random hot deck methods are regarded.

Before the structure of the data sets and the setting is explained, some details on the methods used and evaluated are given in the following.

2.1. Definitions

Since the equivalised household income and the at-risk-of-poverty rate are of crucial importance, they are described in detail in the following.

- *equivalised household income*: It can be seen as the income standardized on a single-person household. Following the description from Statistics Austria on http://www.statistik.at/web_en/statistics/social_statistics/poverty_and_social_inclusion/index.html it "is obtained by dividing the available household income by the number of consumption equivalents in the household. It is assumed that, as the size of the household increases and depending on the age of the children, cost savings are achieved in the household through joint budgeting (economies of scale). For weighting purposes, the EU scale (modified OECD scale) is used to calculate a household's resource requirements. An adult living on his or her own is taken as the reference point (= consumption equivalent), with an allocated weighting of 1. For each additional adult, the assumed resource requirement increases by 0.5 consumption equivalents. Each child under the age of 14 is weighted with a consumption equivalent of 0.3. So a household comprising a father, mother and child would have a calculated consumption equivalent of 1.8 compared with a single-person household."

- *risk-of-poverty*: Again Statistics Austria defines: "The at-risk-of-poverty is calculated on the basis of the equivalised household income. People are considered to be at-risk-of-poverty or affected by the risk of poverty if their equivalised household income is below an at-risk-of-poverty threshold of 60% of the median." (see http://www.statistik.at/web_en/statistics/social_statistics/poverty_and_social_inclusion/index.html).

Of particular interest is the at-risk-of-poverty rate, i.e. the portion of all households that are considered to be at-risk-of-poverty.

So in a mathematical notation the estimation of the at-risk-of-poverty rate from a sample is defined as [see, e.g., Alfons et al., 2013]

$$arpr := \frac{\sum_{i \in I_{<arpt}} w_i}{\sum_{i=1}^{n} w_i} \cdot 100 \quad ,$$

where $I_{<arpt} := \{i \in \{1, \ldots, n\} : x_i < arpt\}$, $\mathbf{x} := (x_1, \ldots, x_n)^T$ with $x_1 \leq \ldots \leq x_n$, is the equivalised household income, $\mathbf{w} := (w_1, \ldots, w_n)^T$ are the corresponding sample weights, n the number of observations and $arpt$ is the estimated at-risk-of-poverty threshold, $arpt = 0.6 \cdot \hat{q}_{0.5}$, where $\hat{q}_{0.5}$ is the weighted median defined as

$$\hat{q}_{0.5} = \hat{q}_{0.5}(\mathbf{x}, \mathbf{w}) = \begin{cases} \frac{1}{2}(x_j + x_{j+1}) & \text{if } \sum_{i=1}^{j} w_i = 0.5 \cdot \sum_{i=1}^{n} w_i \\ x_{j+1} & \text{if } \sum_{i=1}^{j} w_i < 0.5 \cdot \sum_{i=1}^{n} w_i < \sum_{i=1}^{j+1} w_i \end{cases} \quad .$$

The threshold in 2010 was at a equivalised income of 12371 euros per year (or about 1031 euros a month (12 times)) for a single-person household [see Glaser and Heuberger, 2012].

2.2. Distance Measures

Distance measures are of great importance for statistical matching methods. Therefore some popular distance measures get described, because they are mentioned and used later on (see also Chapter 2.4).

Let $\mathbf{x_1}, \mathbf{x_2}, \ldots, \mathbf{x_n}$ be p-dimensional vectors. In general a real-valued distance function d fulfills the following properties [see, e.g., D'Orazio et al., 2006]:

1. for any two vectors $\mathbf{x_k}$ and $\mathbf{x_l}$ the function d is symmetric, i.e. $d(\mathbf{x_k}, \mathbf{x_l}) = d(\mathbf{x_l}, \mathbf{x_k})$,

2. for any two vectors $\mathbf{x_k}$ and $\mathbf{x_l}$ the function d is nonnegative, i.e. $d(\mathbf{x_k}, \mathbf{x_l}) \geq 0$, and

3. for any vector $\mathbf{x_k}$ holds, that $d(\mathbf{x_k}, \mathbf{x_k}) = 0$ (property of identity).

Looking at a data set $\mathbf{X}$ with n observations and p variables for each observation, in order to compute the distances it is necessary to distinguish between the several types of the variables, i.e. to consider if the variables are continuous, categorical (maybe binary) or semi-continuous.

4

Minkowsky distance

A common class of distance measures is based on the Minkowsky distance, which is defined in D'Orazio et al. [2006] as

$$d(\mathbf{x_k}, \mathbf{x_l}) = \left[\sum_{j=1}^{p} c_j^{\lambda} |x_{kj} - x_{lj}|^{\lambda} \right]^{\frac{1}{\lambda}} \quad ,$$

with $\lambda \geq 1$ and c_j a scaling factor for the jth entry. Note that the Minkowsky distance is often defined without the scaling factor c_j, that is $c_j = 1$. It is used for continuous variables.

Manhattan distance

A representative of the Minkowsky distance is for example the Manhattan distance. The parameter λ is defined as 1 and this distance function looks like [see, e.g., D'Orazio et al., 2006]

$$d(\mathbf{x_k}, \mathbf{x_l}) = \sum_{j=1}^{p} c_j |x_{kj} - x_{lj}| \quad .$$

Euclidean distance

Another distance function based on the Minkowsky distance is the Euclidean distance. λ is set to 2, so the distance function is defined as [see, e.g., D'Orazio et al., 2006]

$$d(\mathbf{x_k}, \mathbf{x_l}) = \sqrt{\sum_{j=1}^{p} c_j^2 (x_{kj} - x_{lj})^2} \quad .$$

Maximum distance

The last mentioned representative of the Minkowsky distance is the maximum distance, also called Chebyshev distance. It results letting λ converge to infinity and therefore looks like [see, e.g., D'Orazio et al., 2006]

$$d(\mathbf{x_k}, \mathbf{x_l}) = \max_{j \in \{1, \ldots, p\}} \{ c_j |x_{kj} - x_{lj}| \} \quad .$$

Mahalanobis distance

A distance measure including the covariance matrix of the vectors, $\mathbf{\Sigma_{XX}}$, and also used for continuous data, is the Mahalanobis distance. It is defined as [see, e.g., D'Orazio et al., 2006]

$$d(\mathbf{x_k}, \mathbf{x_l}) = (\mathbf{x_k} - \mathbf{x_l})' \mathbf{\Sigma_{XX}}^{-1} (\mathbf{x_k} - \mathbf{x_l}) \quad .$$

Gower distance

An adequate distance measure for mixed type variables is the Gower distance. It looks like [see, e.g., D'Orazio et al., 2006]

$$d(\mathbf{x_k}, \mathbf{x_l}) = \frac{1}{p} \sum_{j=1}^{p} c_j d(x_{kj}, x_{lj}) \quad ,$$

where c_j is set to 1 for binary variables and $c_j = \frac{1}{R_j}$, with R_j defined as the range of the jth variable, $R_j = \max_k\{x_{kj}\} - \min_k\{x_{kj}\}$, for continuous and categorical ordinal variables. $d(x_{kj}, x_{lj})$ is usually defined as $|x_{kj} - x_{lj}|$, see D'Orazio et al. [2006].

2.3. Regression Models Including Selected Small Area Methods

2.3.1. Linear Regression

Regression analysis is used to predict some values of one or more so-called dependent variables with known (independent) variables. In linear regession a linear relationship between the independent and dependent variables is tried to be found.

In the rest of the chapter, only the case of one dependent variable is described.

Multiple Linear Regression

Multiple linear regression means that one dependent variable $\mathbf{Y}$, the response, is tried to get explained with one or more (independent) predictor variables $\mathbf{x_1}, \dots, \mathbf{x_q}$. $q + 1$ is equal to the number of variables in the model matrix (see also the next page for further explanations). For instance the dependent variable Y could be the income and the independent variables could be the age, the state, the highest completed level of education and the occupational status.

In the multiple linear model $\mathbf{Y}$ is a linear combination of the $\mathbf{x_1}, \dots, \mathbf{x_q}$ including maybe a constant term, referred to as the intercept, and a random error $\boldsymbol{\varepsilon}$. The values of the predictors $\mathbf{x_1}, \dots, \mathbf{x_q}$ are fixed. The error $\boldsymbol{\varepsilon}$ includes the influence of other latent variables missing in the model. It is treated as a random variable with special properties, hence also $\mathbf{Y}$ is random. So now given n independent observations of Y and the associated $x_1, \dots, x_q$ the linear regression model looks like [see, e.g., Johnson and Wichern, 1998]

$$Y_1 = \beta_0 + \beta_1 x_{11} + \dots + \beta_q x_{1q} + \varepsilon_1$$
$$Y_2 = \beta_0 + \beta_1 x_{21} + \dots + \beta_q x_{2q} + \varepsilon_2$$
$$\vdots$$
$$Y_n = \beta_0 + \beta_1 x_{n1} + \dots + \beta_q x_{nq} + \varepsilon_n \quad,$$

or using matrix notation

$$\mathbf{Y} = \mathbf{X}\boldsymbol{\beta} + \boldsymbol{\varepsilon} \quad, \tag{1}$$

whereas $\mathbf{Y}$ is a random vector of dimension n, $\mathbf{X}$ the matrix of dimension $n \times (q + 1)$ with ones in the first column, $\boldsymbol{\beta}$ the $(q + 1)$-vector $(\beta_0, \beta_1, \dots, \beta_q)$ and $\boldsymbol{\varepsilon}$ a random vector of dimension n. In the following $\mathbf{Y}$ stands for the random vector and $\mathbf{y}$ stands for the vector with concrete realisations.

There are some assumptions to the random vector $\boldsymbol{\varepsilon}$: The expected value of ε_i is 0 $\forall i = 1, \dots, n$, the variance of the errors is equal and constant $\forall i = 1, \dots, n$ (this is called "homoscedasticity") and the errors are uncorrelated. This can be written as [see, e.g.,

6

Johnson and Wichern, 1998]

1. $\mathbb{E}(\boldsymbol{\varepsilon}) = \mathbf{0}$ and

2. $Cov(\boldsymbol{\varepsilon}) = \mathbb{E}(\boldsymbol{\varepsilon}\boldsymbol{\varepsilon}^T) = \sigma^2 \mathbf{I}_n$

$$(2)$$

In case of ordinary least squares (OLS) the normal assumption of $(\mathbf{Y}, \mathbf{X})$ is necessary to test hypothesis or to estimate confidence intervals when using the classical method.

Ordinary Least Squares (OLS) Estimation

In general, the parameters $\boldsymbol{\beta}$ (i.e. $\beta_0, \beta_1, \ldots, \beta_q$) and σ^2 are unknown and they have to be estimated to predict the response with given predictor variables. There are lots of possibilities to perform the estimation. One of the best known methods is the Ordinary Least Quares (in the following also denoted by OLS). The idea is to minimize the sum of the squared differences $y_i - \beta_0 - \beta_1 x_{i1} - \ldots - \beta_q x_{iq}$. These differences are called residuals and are denoted by ε_i.

$$\sum_{i=1}^{n} (y_i - \beta_0 - \beta_1 x_{i1} - \ldots - \beta_q x_{iq})^2 = (\mathbf{y} - \mathbf{X}\boldsymbol{\beta})'(\mathbf{y} - \mathbf{X}\boldsymbol{\beta}) = \boldsymbol{\varepsilon}'\boldsymbol{\varepsilon} \to \min$$

The obtained OLS estimates of $\boldsymbol{\beta}$ is denoted by $\hat{\boldsymbol{\beta}}$ and it is expressed by the matrix multiplication $\hat{\boldsymbol{\beta}} = (\mathbf{X}'\mathbf{X})^{-1}\mathbf{X}'\mathbf{y}$ [see, e.g., Johnson and Wichern, 1998]. If $\mathbf{X}$ hasn't full rank $q + 1$, for the inverse $(\mathbf{X}'\mathbf{X})^{-1}$ a generalized inverse of $\mathbf{X}'\mathbf{X}$ is used. The so-called fitted values of $\mathbf{y}$ are given as the estimation of $\mathbf{y}$: $\hat{\mathbf{y}} = \mathbf{X}\hat{\boldsymbol{\beta}} = \mathbf{X}(\mathbf{X}'\mathbf{X})^{-1}\mathbf{X}'\mathbf{y}$. Also the estimated residuals can be calculated using only simple matrix operations of $\mathbf{X}$ and $\mathbf{y}$, $\hat{\boldsymbol{\varepsilon}} = \mathbf{y} - \hat{\mathbf{y}} = (\mathbf{I}_n - \mathbf{X}(\mathbf{X}'\mathbf{X})^{-1}\mathbf{X}')\mathbf{y}$ [see, e.g., Johnson and Wichern, 1998].

In this context it is important to declare a model matrix. It has ones in the first column for the intercept. The rest of the matrix is built by the predictor variables. In the case of a continuous variable the corresponding column of the model matrix contains simply the several values of the observations. In the case of a factor variable the group membership is distinguished by more than one column in the model matrix. So in the case of k groups for a factor, the model matrix gets added $k - 1$ columns, whereas the first group is the reference group. If an observation belongs to the first group, the $k - 1$ added columns will have the value 0 (influence included in the intercept) and if an observation belongs to one of the other $k - 1$ groups, the corresponding column will have a 1 as entry and the remaining columns again 0. [see, e.g., Sachs and Hedderich, 2009]

Quality measure

A common measure for the quality of the estimation is the coefficient of determination

R^2, which is given as [see, e.g., Johnson and Wichern, 1998]

$$R^2 = 1 - \frac{\sum_{j=1}^{n} \hat{\varepsilon}_j^2}{\sum_{j=1}^{n} (y_j - \overline{y})^2} = \frac{\sum_{j=1}^{n} (\hat{y}_j - \overline{y})^2}{\sum_{j=1}^{n} (y_j - \overline{y})^2}$$

whereas $\overline{y}$ is the arithmetic mean of the y_i, so $\overline{y} = \frac{1}{n} \sum_{i=1}^{n} y_i$. It describes the (by the predictor variables) explained proportion of the total variance of the y_i. R^2 takes values from 0 to 1. A value of 1 means that all estimated residuals are 0 (perfect linear relationship) and on the other side a value of 0 means, that all regression coefficients except the intercept β_0 are 0, so the predictor variables $x_1, \ldots, x_q$ have no bearing on the response (no linear relationship). Generally the higher the value, the better the fit. But caution is required, because sometimes it leads to false conclusions. With increasing number of predictor variables also R^2 takes a higher value. In this context the so-called adjusted R-squared has to be mentioned, the number of predictor variables are taken into account here [see, e.g., Sachs and Hedderich, 2009].

Some properties

The estimations $\hat{\boldsymbol{\beta}}$ and $\hat{\boldsymbol{\varepsilon}}$ of the classical regression model (1) with the assumptions (2) have some desirable properties [see, e.g., Johnson and Wichern, 1998]:

1. $\mathbb{E}(\hat{\boldsymbol{\beta}}) = \boldsymbol{\beta}$ and $Cov(\hat{\boldsymbol{\beta}}) = \sigma^2 (\mathbf{X}'\mathbf{X})^{-1}$

2. $\mathbb{E}(\hat{\boldsymbol{\varepsilon}}) = (0)$ and $Cov(\hat{\boldsymbol{\varepsilon}}) = \sigma^2 (\mathbf{I}_n - \mathbf{X}(\mathbf{X}'\mathbf{X})^{-1}\mathbf{X}')$

3. $\hat{\boldsymbol{\beta}}$ and $\hat{\boldsymbol{\varepsilon}}$ are uncorrelated.

4. The OLS-fit of $\hat{\boldsymbol{\beta}}$ is the best linear unbiased estimator (BLUE)(when $\mathbf{X}$ has full rank).

Normal distributed errors

Under the assumption of normal distribution of the residuals $\boldsymbol{\varepsilon}$ with mean vector $\mathbf{0}$ and covariance matrix $\sigma^2 \mathbf{I}_n$, e.g. $\boldsymbol{\varepsilon} \sim \mathcal{N}(0, \sigma^2 \mathbf{I}_n)$, in case of the classical linear regression model (1) where $\mathbf{X}$ has full rank $q + 1$, the OLS estimator $\hat{\boldsymbol{\beta}}$ is distributed as $\mathcal{N}(\boldsymbol{\beta}, \sigma^2 (\mathbf{X}'\mathbf{X})^{-1})$ [see, e.g., Johnson and Wichern, 1998].

Weighted linear regression (WLS)

In the case of heterogeneity of variance of the residuals (i.e. if the second assumption of (2) of the classical linear regression model is violated), the OLS estimator will no longer be the "BLUE", because there is loss of efficiency [see, e.g., Heeringa et al., 2010]. But there is the possibility to put things right if "weights" will be used. The properties

1. $\mathbb{E}(\varepsilon) = \mathbf{0}$ and
2. $Cov(\varepsilon) = \sigma^2 \Sigma$,

whereas Σ is not the identity matrix, but a $n \times n$ diagonal matrix, are considered. The diagonal of $\sigma^2 \Sigma$ contains the reciprocal values of the weights, what are the residuals variances. A transformation of the original model

$$\mathbf{Y} = \mathbf{X}\beta + \varepsilon$$

to

$$(\Sigma^{-1/2}\mathbf{Y}) = (\Sigma^{-1/2}\mathbf{X})\beta + (\Sigma^{-1/2}\varepsilon)$$

results in a model that fulfills the "classical" assumptions (2), because $Cov(\Sigma^{-1/2}\varepsilon) = \Sigma^{-1/2}Cov(\varepsilon)\Sigma^{-1/2} = \Sigma^{-1/2}\sigma^2\Sigma\Sigma^{-1/2} = \sigma^2\mathbf{I}_n$ [see, e.g., Heeringa et al., 2010]. So the WLS estimator, which is at the same time the OLS estimator of this transformed model, is BLUE again and looks like [see, e.g., Heeringa et al., 2010]

$$\hat{\beta} = ((\Sigma^{-1/2}\mathbf{X})'\Sigma^{-1/2}\mathbf{X})^{-1}(\Sigma^{-1/2}\mathbf{X})'\Sigma^{-1/2}\mathbf{y} =$$
$$= (\mathbf{X}'(\Sigma^{-1/2})'\Sigma^{-1/2}\mathbf{X})^{-1}\mathbf{X}'(\Sigma^{-1/2})'\Sigma^{-1/2}\mathbf{y} =$$
$$= (\mathbf{X}'\Sigma^{-1}\mathbf{X})^{-1}\mathbf{X}'\Sigma^{-1}\mathbf{y} \quad .$$

So with the denotation of $\mathbf{W}$ for the matrix of the weights, that is $\mathbf{W} = \sigma^2\Sigma^{-1}$, the WLS estimator is written as $(\mathbf{X}'\mathbf{W}\mathbf{X})^{-1}\mathbf{X}'\mathbf{W}\mathbf{y}$ [see, e.g., Heeringa et al., 2010].

Working with complex sample survey data it is necessary to use weights in order to include the differences in sample inclusion probabilities, unit nonresponse and so on. The weight of a unit specifies the number of people that are represented by the particular observation. These sampling weights can be integrated in the regression model via the WLS estimation. So the weighted least squares estimation results in the formula $\hat{\beta} = (\mathbf{X}'\mathbf{W}\mathbf{X})^{-1}\mathbf{X}'\mathbf{W}\mathbf{y}$ for the regression parameters, where $\mathbf{W}$ is the diagonal matrix of the weights.

2.3.2. Robust Linear Regression

By reason of the high sensibility to outliers of the linear regression method, the interest is in methods that can deal properly with data containing outliers. The so called "breakdown point" is a measure for the robustness, which indicates the minimum proportion of the data contaminated with outliers that can make the estimator "useless" (e.g. the estimator takes arbitrarily large values) [see, e.g., Rousseeuw and Leroy, 2003]. OLS estimators has a breakdown point of $\frac{1}{n}$. $\frac{1}{n}$ converges to zero for increasing sample size n and hence it can be said that OLS estimation has a breakdown point of 0%.

There are several methods to increase the breakdown point of regression estimators, but in this contribution only the idea of so called "M"-estimators is explained. Instead of minimizing the sum of the squared residuals, the aim is to minimize the sum of another function ρ of the residuals [see, e.g., Rousseeuw and Leroy, 2003],

$$\sum_{i=1}^{n} \rho(y_i - \beta_0 - \beta_1 x_{i1} - \ldots - \beta_q x_{iq}) = \sum_{i=1}^{n} \rho(\varepsilon_i) \to \min$$

For logical reasons this function ρ is a symmetric function, i.e. $\rho(t) = \rho(-t) \; \forall t$, and it has a unique minimum at zero [see, e.g., Rousseeuw and Leroy, 2003]. A well known version of M-estimators is Huber's M-estimator, where the equation

$$\sum_{i=1}^{n} \min(c, \max(\frac{\varepsilon_i}{\hat{\sigma}}, -c))\mathbf{x_i} = 0$$

with $\mathbf{x_i} = (x_{i1}, \ldots, x_{ip})$ and $\mathbf{0} = (0, \ldots 0)$ has to be solved. Details can be found in the book "Robust Regression and Outlier Detection" of Rousseeuw and Leroy [2003].

2.3.3. Logistic Regression

The intention is to model and predict a binary variable. Due to the fact that linear regression models require a continuous response variable, they can't be used for the estimation. A popular technique to consider binary responses is logistic regression.

The solution is to work with the posterior probabilities $\mathbb{P}(Y = 0 | \mathbf{X} = \mathbf{x})$ and $\mathbb{P}(Y = 1 | \mathbf{X} = \mathbf{x})$, so the probability that the response of one observation belongs to category 0 or 1 given the corresponding predictor variables. The idea is that a transformation of the posterior probabilities is linear in $\mathbf{x}$. The model looks like [see, e.g., Hastie et al., 2009]

$$\log \frac{\mathbb{P}(Y = 1 | \mathbf{X} = \mathbf{x})}{\mathbb{P}(Y = 0 | \mathbf{X} = \mathbf{x})} = \beta_0 + \beta_1 x_1 + \ldots + \beta_q x_q \quad .$$

One can show that

$$\mathbb{P}(Y = 1 | \mathbf{X} = \mathbf{x}) = \frac{\exp(\beta_0 + \beta_1 x_1 + \ldots + \beta_q x_q)}{1 + \exp(\beta_0 + \beta_1 x_1 + \ldots + \beta_q x_q)} \quad \text{and}$$

$$\mathbb{P}(Y = 0 | \mathbf{X} = \mathbf{x}) = \frac{1}{1 + \exp(\beta_0 + \beta_1 x_1 + \ldots + \beta_q x_q)} \quad .$$

The used transformation is the monotone logit-transformation: $\text{logit}(p) = \log\left(\frac{p}{1-p}\right)$. Note that the posterior probabilities sum up to 1.

The estimation of the parameter vector β is usually made with the maximum likelihood method and the solving is done iteratively [see, e.g., Hastie et al., 2009]. There is no closed form of the solution as it is the case in linear regression models.

Again it is necessary to work with weights if it is handled with real sample survey data. The weights get incorporated in the logistic regression model in the weighted (pseudo-)(log-)likelihood function used for the calculation of the parameters. For details on the weighted logistic regression see Heeringa et al. [2010].

2.3.4. Linear Mixed Regression

The idea of mixed regression models is, in addition to fixed effects, the inclusion of so called "random effects". While the regression coefficients of the previous sections are now noted as fixed effects, random effects are added to the model. Every unit of a sample survey belongs to a certain domain. A domain is a subset of the population U such as for example the people of a federal state, another geographical area population or also a class defined by age and gender. If it was worked as before only with fixed effects so as to incorporate these differences between the classes or domains, there would be too many parameters that would have to be estimated. So now random effects get added in order that possible differences between the different domains get considered in a model.

A mixed effects model can be formulated as [see, e.g., Christensen, 2011]

$$\mathbf{Y} = \mathbf{X}\boldsymbol{\beta} + \mathbf{Z}\mathbf{u} + \boldsymbol{\varepsilon} \quad ,$$

where $\mathbf{u}$ is a vector of domain-specific random effects $u_d \sim \mathcal{N}(0, \sigma_u^2)$ for domain U_d and $\mathbf{Z}$ a matrix. For a single observation $k \in U_d \subset U$ the model for domain-specific random intercepts is given by

$$Y_k = \mathbf{x}_k'\boldsymbol{\beta} + u_d + \varepsilon_k \quad ,$$

with $\varepsilon_k \sim \mathcal{N}(0, \sigma^2)$.

First, the values of $\boldsymbol{\beta}$, σ_u^2 and σ^2 have to be estimated. Afterwards the values of the random effects are estimated.

For estimated samples from complex surveys, weights should be included in the model. Details can be found in the resources of the AMELI project 2011 [see, e.g., Lehtonen et al., 2011].

2.3.5. Logistic Mixed Regression

Analogous to linear mixed effects regression models, random effects can be included in the logistic mixed effects model as well. Hence the mixed logistic model is given by [see, e.g., Lehtonen et al., 2011]

$$\mathbb{P}(Y_k = 1|u_d) = \frac{\exp(\beta_0 + \beta_1 x_1 + \ldots + \beta_q x_q + u_d)}{1 + \exp(\beta_0 + \beta_1 x_1 + \ldots + \beta_q x_q + u_d)} \quad ,$$

with unit $k \in U_d$, $\boldsymbol{\beta}$ the vector of the fixed effects und u_d the domain-specific random effect.

Again weights can be included in the model working with sample survey data [see Lehtonen et al., 2011].

2.3.6. The Transmission of the Model

The procedure is to use regression methods for the estimation of the variable of interest in sample survey A. For that reason the model observed for sample survey B is transmitted to the sample survey A. For this purpose a regression model is built in the sample survey B using predictor variables that are observed in both sample surveys A and B, and using the variable of interest as response, i.e., regarding for example linear regression, $\hat{\beta}_B = (X_B{}'X_B)^{-1}X_B{}'y_B$ with X_B the model matrix built by intercept and predictor variables from survey B (here: EU-SILC) and y_B the response variable in survey B.

The estimated parameter vector $\hat{\beta}_B$ is now used as parameter vector in the model for A, more precisely, in the model $\hat{y}_A = X_A\hat{\beta}_B$, where in $\hat{y}_A$ are the estimated values for the variable of interest, that is initially missing in A, and X_A is the model matrix of survey A, including ones in the first column and the corresponding interaction and contrasts for the common variables of A and B (used for the model estimation in survey B) in the other columns. Note that the (common) variables in X_A have to be in the same order as in X_B, the model matrix (with ones in the first column) used in the estimated model for survey B.

2.4. Statistical Matching

In general two approaches in statistical matching exists: the micro and the macro approach. In the following only the micro approach is considered. This is the version where in the survey A an additional variable is simulated, i.e. values (of the variable observed in survey B and not observed in survey A) get imputed, so that a "synthetic" data set results. The survey A is called "recipient" file and survey B the "donor" file. [see, e.g., D'Orazio et al., 2006]

Table 1: Survey A and B in one data set: the variables X_j, $j = 1, \ldots, f$, are observed only in survey A, the variables Z_j, $j = 1, \ldots, h$, are observed only in survey B and the variables Y_j, $j = 1, \ldots, g$ are observed both in A and B.

survey	X_1	X_2	...	X_f	Y_1	Y_2	...	Y_g	Z_1	Z_2	...	Z_h
A	x_{11}^A	x_{12}^A	...	x_{1f}^A	y_{11}^A	y_{12}^A	...	y_{1g}^A				
	x_{21}^A	x_{22}^A	...	x_{2f}^A	y_{21}^A	y_{22}^A	...	y_{2g}^A				
	$\vdots$	$\vdots$	...	$\vdots$	$\vdots$	$\vdots$	...	$\vdots$				
	$x_{n_A1}^A$	$x_{n_A2}^A$	...	$x_{n_Af}^A$	$y_{n_A1}^A$	$y_{n_A2}^A$	...	$y_{n_Ag}^A$				
B					y_{11}^B	y_{12}^B	...	y_{1g}^B	z_{11}^B	z_{12}^B	...	z_{1h}^B
					y_{21}^B	y_{22}^B	...	y_{2g}^B	z_{21}^B	z_{22}^B	...	z_{2h}^B
					$\vdots$	$\vdots$	...	$\vdots$	$\vdots$	$\vdots$	...	$\vdots$
					$y_{n_B1}^B$	$y_{n_B2}^B$	...	$y_{n_Bg}^B$	$z_{n_B1}^B$	$z_{n_B2}^B$	...	$z_{n_Bh}^B$

Different possibilities to perform the imputation are available, for example, the non-parametric techniques "nearest neighbor distance hot deck imputation" (in the following also denoted as NND), the "random hot deck imputation" (in the following also denoted as RND), the "sequential hot deck procedure" (in the following also denoted as SEQ)[see, e.g., Madow et al., 1983, Madow and Olkin, 1983] or the "rank hot deck imputation" (in the following also denoted as RNK) [see D'Orazio, 2012].

To explain the methods first the term "matching variables" is described: These are some (or all) of the variables, say Y_j, observed in A as well as in B, usually the most relevant ones [compare Table 1]. They are used for the matching procedure (for the measurement of the distance) and the choice of the matching variables should be based on statistical selection criteria. Some of the methods could be the computation of the pairwise correlation or association measures, Cramer's V, "proportional reduction of the variance" measures, ranks, the (adjusted) R^2 from a regression model and much more. [see also D'Orazio, 2012]

The ideas of the above mentioned methods are as follows [see also D'Orazio, 2012]:

1. NND: Distances between the matching variables of either all or just a subset of units get measured and the closest donor unit is chosen to the given recipient unit. In the case of subsets it is spoken of donation classes. The advantage of them is the lower computational effort. Possible distance measures are - depending on the type of variables (continuous, categorical, semi-continuous) - for example the Manhattan distance, the maximum distance, the Euclidean, the Mahalanobis distance or, if variables of different type are considered, the Gower distance. If there are more than one closest unit, one of them is chosen at random.

2. RND: Here a donor unit is chosen completely at random out of a appropriate subset of units. This subset can be performed in several ways. One possibility is to take all units with the same particular characteristics as the regarded recipient unit (for example the same geographical area and the same gender)(compare "donation classes" explained for NND). Other possibilities can be implemented with the help of the matching variables. For example to pick a unit at random out of the k nearest units, out of units with maximal distance i or out of a proportion j $(0 < j < 1)$ of the closest donors.

3. SEQ: The idea of sequential hot deck is to sort the data on the basis of some defined "ordering variables" before the imputation is done. If classification groups (donation classes) are used, the ordering will happen within each group. Then for every recipient unit, the previous reported unit out of the donor data set is used as donor. For further details on this method see Madow et al. [1983] and Madow and Olkin [1983].

4. RNK: Again distances between the matching variables get computed and the unit with the minimum distance is picked as donor for a given recipient. The distances get computed on the empirical cumulative distribution functions of the single (continuous) considered matching variable. If the estimated cumulative distribution function of the matching variable of the recipient file is denoted with $\hat{F}^A$ and the one of the donor file with $\hat{F}^B$, the unit b^* of B with

$$|\hat{F}^A(x_a) - \hat{F}^B(x_{b^*})| = \min_{1 \leq b \leq n_B} |\hat{F}^A(x_a) - \hat{F}^B(x_b)|$$

will be chosen as donor for the regarded record a of A, whereas x_a (x_b) denotes the value of the matching variable of the unit a (b) of the data set A (B) and n_B the number of records in B. Again donation classes may be used. In this case the distribution function gets estimated for each class separately.

For the methods NND and RNK an alternative is to use a "constrained" method. Generally a unit can be choosen several times as donor. "Constrained" means that every unit is chosen only once as a donor. Here the overall matching distance gets minimized and this overall distance is greater than the overall distance in the unconstrained case, but it may end in better results. A constraint for the constrained method is clearly that the donor file has to be equal or greater than the record file. [see, e.g., D'Orazio, 2012]

An important and interesting approach is to take the (sampling) weights into account. In combination with RND this is done in the way that the donors get picked with propability proportional to the weights corresponding to the donors. On the other hand in combination with RNK the weights w_i are used to calculate the empirical cumulative distribution function:

$$\hat{F}(x) = \frac{\sum_{i=1}^{n} w_i I(x_i \leq x)}{\sum_{i=1}^{n} w_i}$$

with $I(.)$ as the indicator function. [see also D'Orazio, 2012]

In the literature you can find some statistical methods that explicitly include the sampling design and weights (for example an approach with calibration), see D'Orazio [2012] and D'Orazio et al. [2006].

2.5. Bootstrap Methods

Bootstrap methods are replication-based methods for the evaluation of the variance of a quantity calculated on the basis of a data set. Several possibilities to perform bootstrapping exists [see, e.g., Wolter, 2010, Hastie et al., 2009]. In the following one bootstrap method is described.

The basis of the calculation is the original data set, lets say $\mathbf{D}$, a $n \times (q+1)$ matrix with $(Y_i, X_{i1}, \ldots, X_{iq})$ the i-th observation.

Bootstrap samples (also called bootstrap replicates) of size $n^* = n$ (in the original bootstrap method; other sizes, for example $n - 1$, can be used) get drawn of $\mathbf{D}$ with simple random sampling with replacement. The (large) number of bootstrap samples is denoted with R and one bootstrap sample with $\mathbf{D}^{(*r)}$, so there are $\mathbf{D}^{(*1)}$, $\mathbf{D}^{(*2)}$, ..., $\mathbf{D}^{(*R)}$. Every of the $\mathbf{D}^{(*r)}$, $r = 1, \ldots, R$, is of dimension $n^* \times (q+1)$. [see, e.g., Wolter, 2010]

Now the above described methods can be used for each bootstrap sample $\mathbf{D}^{(*r)}$. In detail R bootstrap replicates of the data set A and also R bootstrap samples of the data set B are drawn. In connection with the model transmission methods, the model is estimated for the r-th bootstrap sample of B and it gets transmitted to the r-th bootstrap

14

sample of A. On the other hand, in combination with statistical matching, in the r-th bootstrap sample of A are imputed values received from the r-th bootstrap sample of B.

The quantity of interest $S(\mathbf{D})$ get calculated of every $\mathbf{D}^{(*r)}$, so there result $S(\mathbf{D}^{(*1)}), \ldots$, $S(\mathbf{D}^{(*R)})$ and the distribution of $S(\mathbf{D})$ can be estimated. For example the variance could be of interest [see, e.g., Hastie et al., 2009]:

$$\widehat{Var}(S(\mathbf{D})) = \frac{1}{R-1}\sum_{r=1}^{R}(S(\mathbf{D}^{(*r)}) - \bar{S}^*)^2 \quad \text{with} \quad \bar{S}^* = \frac{1}{R}\sum_{r=1}^{R}S(\mathbf{D}^{(*r)})$$

For this thesis beside of the variance estimation the mean of the bootstrap results, $\bar{S}^*$, is of interest.

An alternative bootstrap method in context with regression would be bootstrapping residuals. The background therefor is that the regressors are treated as fixed and not as random as it is the case if the bootstrapping concerns the observations. It has comparably good properties, the information in the predictor variables gets retained. [see, e.g., Efron and Tibshirani, 1993]

Due to comparability aspects with statistical matching procedures this method is not used and evaluated in this thesis.

Furthermore, the calibrated weights bootstrap method [see, e.g., Alfons and Templ, 2013] is a reasonable alternative and extension to the naive bootstrap approach used in this thesis. "While the naive bootstrap does not modify the weights of the bootstrap samples, a calibrated version allows to calibrate each bootstrap sample on auxiliary information before deriving the bootstrap replicate estimate. [...] While the naive bootstrap is faster to compute, the calibrated bootstrap in general leads to more precise results." [Templ and Alfons, 2013] However, due to the fact, that the method is computationally intensive, it is not used in this thesis.

3. Design of the Simulation Study

3.1. Problem

As already mentioned, the topic of this thesis is the evaluation and comparison of several methods concerning the estimation of regional indicators.

An obvious approach of estimating poverty is to estimate poverty indicators simply on basis of a survey A that includes information on variables, that can be used for the calculation of the indicators (here on the basis of an EU-SILC data set). However, due to the small sample size of EU-SILC, variances can be large when estimating indicators on regional level.

Maybe transmission of information to a larger survey B, in this context a Micro Census data set, can lead to a significant decrease of variances.

3.2. Set-Up of the Design-Based Simulation Study

For the purpose described in Chapter 3.1 a close-to-reality population U has been simulated that contains information on all variables of interest. This population set is the basis and it includes all variables that appear in both sample surveys A and B or in either one of them.

This approach is called design-based simulation, because from the simulated population samples are drawn considering the different sample designs. The other method would be model-based simulation: the samples itself, i.e. the EU-SILC and Micro Census samples, get simulated. This alternative is not as flexible as the design-based variant, because not any design could be used here to draw samples.

The simulation of the population is done in R [R Core Team, 2013] using among others the package simPopulation [see Alfons and Kraft, 2012]. The process is divided in more steps. First the household structure is constituted (by simulating some basic categorical variables that specify the household structure), then the categorical variables get simulated and afterwards the simulation of the (semi-)continuous variables is done. Important functions for this purpose are simStructure, simCategorical and simContinuous, which can be found in the package simPopulation [Alfons and Kraft, 2012]. The weights are included in every step. Details can be found in Alfons and Kraft [2012] and Alfons et al. [2011].

Thus with the information of the EU-SILC and Micro Census data sets the artificial population U is simulated, compare also Figure 1.

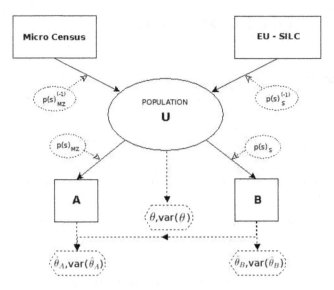

Figure 1: The idea of constructing the artificial samples and evaluating the estimators.

Following the initial denotation, the (artificial) sample survey Micro Census equates sample survey A and the (simulated) sample survey EU-SILC equates sample survey B.

Now the two data sets A and B are repeatedly drawn, say J times, from the population U considering the particular sample designs $p(s)$, see Figure 1. Details on them can be found in the description of the data, see Chapter 4.1. So one sample with the number of records of A and the variables observed in A and analogous another sample with the number of records of B and the variables observed in sample survey B are drawn. The number of housholds sampled for each state for the simulation of data set A and B, respectively, can be seen in Figure 2 and Figure 3, respectively. It has to be noted that the sample sizes for the EU-SILC data in small states such as Burgenland or Vorarlberg is rather small and large variances of estimates can be expected for those states. In comparison, due to the sampling design and absolute number of respondents, the Micro Census data include much more observations as EU-SILC, especially for small states.

Some versions of the above explained methods in Chapter 2.3 and in Chapter 2.4 get applied to every pair of the two data sets. In other words, the variable of interest is imputed for all observations of each simulated data set A using the known values in the corresponding simulated data set B.

External auxiliary information which is not asked in the questionaire, like for example the internal migration balance rate, the age standardized index of the mortality ratio from 25-54 years, the first quartile of the net income or the portion of non-employed

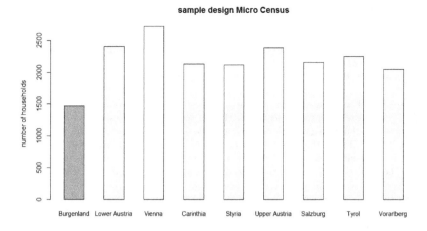

Figure 2: The sample design of Micro Census / data set A.

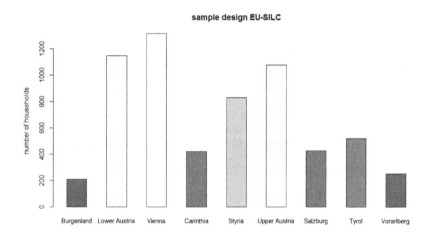

Figure 3: The sample design of EU-SILC / data set B.

19

persons of women/men aged between 30-49 (see Chapter 4.1.3: the described covariates), is partially used.

Because of the knowledge of the "true" values it becomes easy to evaluate and compare the performance of the techniques.
Very important therefor will be some bootstrap methods (see Chapter 2.5).

Assuming that the interest is in a quantity calculated from the (estimated) variable of interest. The "true" quantity is denoted with θ (based on the artificial data set U) and the respective estimation gained from the imputations in the data set A with $\hat{\theta}_A$ and the calculated value of θ for the data set B with $\hat{\theta}_B$ (compare Figure 1). All in all, J values for $\hat{\theta}_A$ and $\hat{\theta}_B$ result, because the data sets A and B get simulated J times as mentioned above. The results are denoted with $\hat{\theta}_{A,j}$ and $\hat{\theta}_{B,j}$ and all these results are stored.

Furthermore for every pair of A and B bootstrap samples, denoted by $A^{(*)}$ and $B^{(*)}$, get drawn. The symbol $^{(*)}$ signifies the bootstrap version. The number of bootstrap replicates let be labelled with R.
To each of the bootstrap sample pair again some versions of the above explained methods in Chapter 2.3 and in Chapter 2.4, namely the same as for A and B, get applied. So the variable of interest is imputed in every bootstrap sample $A^{(*)}$ with the help of the corresponding bootstrap sample $B^{(*)}$.
Again a quantity based on the (estimated) variable of interest is calculated. There result $\hat{\theta}_{A,r}^{(*)}$ and $\hat{\theta}_{B,r}^{(*)}$ with $r = 1, \ldots, R$ for every pair of A and B. The mean and the variance of that values are stored for every A and B.

3.3. Quality Criteria

The comparison and evaluation of the investigated methods is of interest. The aim is to find the "best" method for the estimation of the quantity. Therefore different quality criteria are computed. Interesting are things like the bias, the variance and also the combination of these two quantities, the mean squared error of the estimator, because it is looked for the "best" estimator in matters of small bias and small variance. So the above described repeated drawing of the samples seems to make sense in order to enable an assessment of the bias, variance and MSE of the different methods.

The bias is defined as the difference of the expected value of the estimator and the true value of the quantity [see, e.g., Rinne, 2003]:

$$\text{Bias}(\hat{\theta}) = \mathbb{E}(\hat{\theta}) - \theta \quad . \tag{3}$$

Within our simulation study, $\mathbb{E}(\hat{\theta})$ is replaced with the mean of the parameter estimates $\hat{\theta}_j$, $j = 1, \ldots, J$,

$$\bar{\hat{\theta}} = \frac{1}{J} \sum_{j=1}^{J} \hat{\theta}_j \quad . \tag{4}$$

The variance of the estimator $\hat{\theta}$ is defined as [see, e.g., Rinne, 2003]

$$\text{Var}(\hat{\theta}) = \mathbb{E}[(\hat{\theta} - \mathbb{E}(\hat{\theta}))^2] \quad . \tag{5}$$

Again in our case the variance is replaced with the sample variance [compare, e.g., Rinne, 2003]

$$\frac{1}{J-1} \sum_{j=1}^{J} (\hat{\theta}_j - \bar{\hat{\theta}})^2 \quad . \tag{6}$$

The (sample) standard deviation results by taking the root of the (sample) variance and is denoted with the symbols "SD".

The mean squared error (referred to as MSE) is the expectation of the squared difference of the estimator and the true value of the parameter and it can be written as the sum of the squared bias and the variance of the estimator [see, e.g., Rinne, 2003]:

$$\text{MSE}(\hat{\theta}) = \mathbb{E}[(\hat{\theta} - \theta)^2] = \mathbb{E}[(\hat{\theta} - \mathbb{E}(\hat{\theta}))^2] + (\mathbb{E}(\hat{\theta}) - \theta)^2 = \text{Var}(\hat{\theta}) + \text{Bias}(\hat{\theta})^2. \tag{7}$$

The MSE of an unbiased estimator (i.e. $\text{Bias}(\hat{\theta}) = 0$) is identical with the variance. Again the sample bias and sample variance is used.

Reliability of the Bootstrap

Furthermore, the reliability of the bootstrapping is checked for every method by comparing the standard deviation gained from the repeated sampling of A and B with the standard deviation computed as the mean of the standard deviation received from the bootstrap replications.

$$\text{DSD}_A = \frac{1}{J} \sum_{j=1}^{J} \widehat{\text{SD}}(\hat{\theta}_A^{(*)})_j - \widehat{\text{SD}}(\hat{\theta}_A) \quad , \tag{8}$$

where $\widehat{\text{SD}}(\hat{\theta}_A^{(*)})_j = \sqrt{\frac{1}{R-1} \sum_{r=1}^{R} (\hat{\theta}_{Aj,r}^{(*)} - \bar{\hat{\theta}}_{Aj}^{(*)})^2}$ with $\hat{\theta}_{Aj,r}^{(*)}$, $r = 1, \ldots, R$, being the R bootstrap estimations for the jth sample of A. The formula for sample B is analogously,

$$\text{DSD}_B = \frac{1}{J} \sum_{j=1}^{J} \widehat{\text{SD}}(\hat{\theta}_B^{(*)})_j - \widehat{\text{SD}}(\hat{\theta}_B) \quad , \tag{9}$$

with $\widehat{\text{SD}}(\hat{\theta}_B^{(*)})_j = \sqrt{\frac{1}{R-1} \sum_{r=1}^{R} (\hat{\theta}_{Bj,r}^{(*)} - \bar{\hat{\theta}}_{Bj}^{(*)})^2}$.

In addition the difference in the mean of the at-risk-of-poverty rates from the bootstrap replicates (of all drawn samples) and the mean of the at-risk-of-poverty rates estimated on the basis of the results of the repeated sample drawing will be computed and evaluated. So

$$\text{DM}_A = \frac{1}{J} \frac{1}{R} \sum_{j=1}^{J} \sum_{r=1}^{R} \hat{\theta}_{Aj,r}^{(*)} - \frac{1}{J} \sum_{j=1}^{J} \hat{\theta}_{A,j} \tag{10}$$

is computed for sample A and analogously for sample B

$$\mathrm{DM}_B = \frac{1}{J}\frac{1}{R}\sum_{j=1}^{J}\sum_{r=1}^{R}\hat{\theta}_{Bj,r}^{(*)} - \frac{1}{J}\sum_{j=1}^{J}\hat{\theta}_{B,j} \tag{11}$$

is considered.

Moreover the mean of the bootstrap replications is computed and used for the computation of the bias and the MSE in order to receive in a sense a bias corrected version.

$$\mathrm{Bias}^{(*)}(\hat{\theta}_A^{(*)}) = \frac{1}{J}\frac{1}{R}\sum_{j=1}^{J}\sum_{r=1}^{R}\hat{\theta}_{Aj,r}^{(*)} - \theta \tag{12}$$

$$\mathrm{Bias}^{(*)}(\hat{\theta}_B^{(*)}) = \frac{1}{J}\frac{1}{R}\sum_{j=1}^{J}\sum_{r=1}^{R}\hat{\theta}_{Bj,r}^{(*)} - \theta \tag{13}$$

4. Application to Poverty Estimation Using EU-SILC and Micro Census Data

4.1. Data

The background of these thoughts is the following: Considering the sample surveys Micro Census and EU-SILC, one can detect a large difference regarding the number of observations, the sample fractions in strata and the number of variables. In the Micro Census sample survey much more people are asked, more precisely 23000 households and about 40000 persons get interviewed every quarter. The EU-SILC data set has "only" about 6000 households and 15000 persons. A rough overview of the sampling fractions can be obtained with Figure 2 and Figure 3. As previously mentioned, Burgenland, for example, include much more observations in the Micro Census data set than in the EU-SILC data set. The Micro Census data set doesn't include the variable "at-risk-of-poverty" (for explanation see Chapter 4.1.3) or something equivalent, such as income. However, such a variable is needed to estimate the at-risk-of-poverty rate. The EU-SILC data set includes this information. The idea for gaining information generally, is to gain information in the greater and therefore lets say "better" data set Micro Census (more observations give more information, especially on regions with small sample fraction) and hence to combine both data sets to the effect that such a variable gets estimated with the above-mentioned and explained methods in the Micro Census data set.

In this thesis the indicator calculated from the (for the Micro Census imputed) variable of interest (compare Chapter 3.2) is the at-risk-of-poverty rate (see Chapter 2.1). It gets calculated either with the variable at-risk-of-poverty or on the basis of the equivalised household income (for explanation see again Chapter 2.1). This variable is denoted with *arpr*.

4.1.1. Details on Micro Census

The Micro Census gives information on the employment and unemployment as well as on housing stock and housing conditions.

The sample design is quite simple. It is a stratified (by 9 regions, i.e. states of Austria) simple random sample with (almost) equal size in each strata. Only Vienna has a higher und Burgenland has a lower sample size. The total sample size per quarter is about 23000 households. The samples are drawn out of all private households recorded in the central population register.

All persons (mainly aged 15 years or older) of a sampled household get interviewed 4 times a year (it is a panel design). The participation is compulsory.
The interviewing is operated with CAPI- (Computer Assisted Personal Interviewing

- first interview) and CATI- (Computer Assisted Telephone Interviewing - the following interviews) techniques, respectively.

More details on the Micro Census is given in Moser et al. [2013].

4.1.2. Details on EU-SILC

EU-SILC is the abbreviation for "European Statistics on Income and Living Conditions" and it gives information on people and private households in Europe. The aim is to obtain comparable data on the social position and the income for all EU countries. On the European Union scale it is widely used to report on poverty and social inclusion.

In Austria it was carried out the first time in 2003 by Statistics Austria. It is an annual sample survey where about three-fourths of the households get interviewed again in the following year and one fourth is new, so (since 2004) it is a cross-sectional and longitudinal data collection (thus 2007 the first complete longitudinal sample (over 4 years) was available).

The sample gets drawn from the central population register und its minimum size is 4500 housholds. For example, in 2010 the sample involved 14085 persons in 6188 households.

The deepest used regional stratification is NUTS 2. NUTS is a common classification of territorial units for statistics based on Regulation 1059/2003 of the European Parliament and of the Council. The level NUTS 2 corresponds to the 9 states of Austria.

The sample sizes for the several states are different (proportional to the inhabitants), therefore it is approximately a self-weighting design.

All persons of a household aged 16 years or older get interviewed personally and furthermore some information on children gets collected. The participation is voluntary.

The interviewing is operated with CAPI- and CATI-techniques, respectively.

More details on EU-SILC is given in Glaser and Heuberger [2012].

4.1.3. Variables in the Population

In this section the variables of **U** (the population) used for the estimations get described. Concerning the choice of these variables the decision is based on Bauer et al. [2013] (Statistics Austria).

The estimated variables are the binary variable at-risk-of-poverty and the equivalised household income. The definitions of these two variables can be found in Chapter 2.1.

The used breakdown variable is the state. This means that the analyses are done seperately for every domain, so splitted by the factor state. All the variables are considered at household levels. For the definitions of the variables see also Bauer et al. [2013]. The notation of the variables used later on in the models and the R-Code is given in brackets.

- at-risk-of-poverty (ARPT60i): This variable has the values 0 if there is no risk of poverty and 1, if the persons of the household are at-risk-of-poverty.

- equivalised household income (EQ_INC): This variable is continuous and contains the calculated equivalised household income.

- state (strat): This factor variable indicates the households state of Austria, so there are 9 several levels. (originally NUTS 2 code)

- household type (htyp): Some different types of households are considered, namely single men, single women, multiperson households without children, single parents, multiperson households with one child, multiperson households with two children, multiperson households with three or more children. These 7 levels of the factor variable are coded with 4 to 10, in the same order as they are listed in the previous phrase. (Somebody is classified as a child, if the person is under 16 years old or if the person is under 27 years old, lives in the same household with at least one parent and is not employed.)

- foreign origin of the household (migration): It get distinguished between three factors: 1 for households where at least one person aged 16 years or older has a non-EU nationality or is born in a non-EU country, 2 for households that do not fulfill the conditions for 1, but at least one person aged 16 years or older is born in an EU foreign country or has an EU citizenship and 3 for households where all persons aged 16 and above are Austrian nationals and are born in Austria.

- work intensity of the household (workint): First the number of people of a household that are capable of work gets determined. That are people aged between 18 and 59 years, except a) military and civil service personnel and b) people from 18 to 24 years in education. The work intensity of a household is the mean of the work intensity of all capable of work household members. The work intensity of a person who is fit for work is the ratio of the number of working hours per week and 40. If a person has more than 40 working hours per week, his work intensity is set to 1. The factor variable is divided in 6 levels: -3 for households without persons who are capable of work, 0 for 0% (no work intensity), 1 for $0-25\%$, 2 for $25-50\%$, 3 for $50-75\%$ and 4 for 75% or more.

- population density (urb): The factor is divided into populous (this corresponds to Vienna)(level 1) and other areas (level 0).

- legal relationship of the dwelling (dwell): The several levels of this factor variable first distinguish between for valuable consideration (rent, sublease) and free of charge (freehold, rentfree, related/related by marriage with the owner). Furthermore there is the differentiation in aided flats and other (if for valuable consideration) respectively owner and other (if free of charge). The denotation is 1 for house or apartment owner, 2 for other free of charge legal relationship of the dwelling, 3 for council home (technical term; "Gemeindewohnung") or cooperative flat (technical term; "Genossenschaftswohnung") and 4 for main rental and other for valuable consideration legal relationship.

- Category of the equipment of the apartment (equipm): The level 1 of this factor variable stands for dwelling with bath/shower, WC and central heating, level 2 for the category with bath/shower, WC and single stove heating, 3 for the category with WC and waterconnection inside and 4 for dwellings with only water or without installation.

In addition to pure household characteristics some characteristics evaluated at personal level are of interest. They get built on household level with the help of a reference person. As reference person the person with the highest estimated income is elected. For this person the following characteristics get considered:

- professional branch (rp_branch): This factor variable is splitted into 9 levels and describes the actual (if employed) or former (if unemployed) sector. If the reference person has never been employed it is coded with -3, otherwise the coding is 1 for agriculture and forestry, 2 for manufacturing sector without building industry, 3 for the building industry, 4 for trade, 5 for the accomodation and restaurant industry, 6 for the credit and insurance industry, 7 for public administration, public health and educational system and 8 for other services.

- occupational status (rp_occstat): This factor variable with 8 different levels characterizes the actual (if employed) or former (if unemployed) occupation. Again persons who have never been employed are coded with -3. The other levels are summarized as 1 for manual: unskilled or semi-skilled labour (technical terms; "Hilfs- oder angelernte Arbeit"), 2 for manual: skilled labour, preparatory work, foreman/-master (technical terms; "Facharbeit", "Vorarbeit", "Meister"), 11 for non-manual: unskilled labour (technical term; "Hilfsarbeit"), 12 for non-manual: average skilled occupation (technical term; "mittlere (gelernte) Tätigkeit"), 13 for non-manual: higher, highly qualified, leading occupation, 21 for self-employed/assisting in agriculture and forestry and 22 for self-employed/assisting outside of agriculture and forestry.

- living (rp_living): The three-category factor obtaines the level 1 for unemployed persons who are no retirees, level 2 for retirees and level 3 for employees.

- highest completed level of education (rp_heduc): This factor has 4 levels and is coded as follows: 1 for compulsory school (inclusive without compulsory school graduation), 2 for apprenticeship or vocational school, 3 for grammar school and vocational school with higher education entrance qualification (inclusive course of lectures) and 4 for university and college of higher education.

- family status (rp_famst): For the factor the level 1 stands for unmarried, 2 for married, 3 for widowed and 4 stands for divorced.

- age groups (rp_age): This variable has 7 different levels, namely 10 for reference persons aged younger than 20, 20 for persons from $20 - 29$ years, 30 for persons from $30 - 39$, 40 for persons from $40 - 49$, 50 for persons from $50 - 59$, 60 for persons from $60 - 69$ and 70 for persons aged 70 or older.

Furthermore in some models covariates are added to the so far listed variables. The used covariates get calculated at federal state level and are in detail as follows:

- internal migration balance rate (`migrbal`): This numeric variable is the difference between the internal moving in und the internal departure, standardized to 1000 of the population of the respective region. A value > 0 signifies internal migration gains and a value < 0 internal migration loss.

- portion of non-employed persons of men aged between $30 - 49$ (`nonempm`): This is the rate (in %) of the persons, who are neither employed nor unemployed, but "out of labor force", in which students are excluded from the calculation.

- portion of non-employed persons of women aged between $30 - 49$ (`nonempw`): The continuous variable is defined as `nonempm`, but for women instead of men.

- portion of the live births with a birth weight below 2500 gram (`birth`): This covariate is expressed as a percentage. Note that multiple births are excluded from the calculation.

- age standardized index of the mortality ratio from $25 - 54$ years (`mort`): This variable gets calculated as the ratio of the number of observed deaths of persons prime-aged between 25 and 54 and the expected number of such deaths. The expected value results from the population size of this age group (composed of gender und five-year age groups) and the death rate for Austria. It is an index value, i.e. a value of 116 means, that in a region has been recorded 16% more deaths than would have been expected in Austria with same death rate.

- first quartile (25%) of the net income (in EUR)(`quinc`): This value seperates the 25% lowest-income income recipients of a region from the other 75%. In this context income recipients are defined as wage earners, self-employed persons and retirees. Here income contains beside the earned and self-employment net income also pension and some transfer payments as unemployment benefit and social benefits (technical term; "Notstandshilfe").

- portion of cars with small cylinder capacity (`carscc`): In details it is the portion of the passenger cars/estate cars in the cylinder capacity class lower than 1250 ccm (rate (in %) of all passanger/estate cars). Note that electromotor-actuated cars or cars powered by fuel cells are excluded from the calculation.

- unemployment rate (`unemp`): It is expressed as a percentage.

- extramural social assistance recipients (technical term; "Bezieher von offener Sozialhilfe") respectively means-tested guaranteed minimum income recipients (technical term; "Bezieher von bedarfsorientierter Mindestsicherung") (`socass`): Regarded is the annual sum per 1000 of the population in private households.

- recipients of compensatory allowances (technical term; "Ausgleichszulagenbezieher") in % of the pension stock (`compall`): Notice, that the number of pensions is not the same as the number of retirees and that officials are excluded.

- `unemp_lag`: just like `unemp`, but in relation to the previous year

- `socass_lag`: just like `socass`, but in relation to the previous year

- `compall_lag`: just like `compall`, but in relation to the previous year

4.2. The Models

Now some models with the methods described in Chapter 2 are implemented. For details on the programming code just have a look at the Appendix A.1 R-Code.

4.2.1. Regression Models Including Selected Small Area Methods

Due to the fact that the aim is to estimate the at-risk-of-poverty rate, which is estimated with the equivalised household income or with the binary variable at-risk-of-poverty, two possibilities exist:

1. A model on sample B (equates to EU-SILC) where the response is the equivalised household income is fitted and the model is applied to sample A (equates to the Micro Census) to estimate the needed variable for estimating the at-risk-of-poverty rate. This version corresponds to linear regression models.
   Note:

   - Because (the equivalised household) income is right-skewed, a transformation of the equivalised household income is used to make it more symmetric: it is transformed by the logarithm function as is usually the case.

   - Due to the fact that in the data some zeros for the variable EQ_INC occur, these entries are replaced with ones. So there is no definition problem in connection with the logarithm transformation.

2. First the at-risk-of-poverty is calculated, a model where the dependent variable is the dichotomous variable at-risk-of-poverty is constructed and this model gets transmitted. From the imputed values of the variable at-risk-of-poverty the at-risk-of-poverty rate gets estimated. This variation corresponds to the logistic regression models.

Because of the large number of available variables in the data sets, huge amount of different models can be fitted. Several models using the above described variables are considered (in Chapter 4.1.3). The models described in the following are noted in "R notation". On the left side (left to ∼) the response is written and on the right side (right to ∼) the predictor variables are listed, separated by "+" signs.

(a) **Linear Regression** (compare Chapter 2.3.1)

(i) First a model with the response variable equivalised household income EQ_INC and the predictor variables rp_age, dwell, htyp, migration, equipm, rp_living, rp_famst, workint, rp_heduc, rp_occstat, rp_branch and urb is considered. In the process some of the factors, namely rp_age, equipm, workint and rp_heduc, are used as ordered. The regression is calculated taking into account the weights. So the linear regression model for the income looks like:

$$\log(\text{EQ\_INC}) \sim \text{rp\_age} + \text{dwell} + \text{htyp} + \text{migration} + \text{equipm}$$
$$+ \text{rp\_living} + \text{rp\_famst} + \text{workint} + \text{rp\_heduc} \qquad (14)$$
$$+ \text{rp\_occstat} + \text{rp\_branch} + \text{urb}$$

The model gets estimated for the simulated EU-SILC data set. Then the predictions for the simulated Micro Census data set are estimated and re-transformed in the following with the exponential function. Then the at-risk-of-poverty rate is estimated with these values, on the one hand for Austria and on the other hand also for every of the nine provinces.

(ii) In the second model some of the above mentioned covariates are added as predictor variables to this model. Everything else, meaning the other predictor variables, the use of some ordered factors and the inclusion of weights, remains the same as in the first model. The decision regarding the choice of this covariates is based on an in-depth analysis. Determining factors are the significance of the covariates in the model, the linear dependence of the variables (singularity) and my subjective assessment, which covariates have more or less influence on income and at-risk-of-poverty. So the following model, estimated for the simulated EU-SILC data set and transmitted to the artificial Micro Census data set, results:

$$\log(\text{EQ\_INC}) \sim \text{rp\_age} + \text{dwell} + \text{htyp} + \text{migration} + \text{equipm}$$
$$+ \text{rp\_living} + \text{rp\_famst} + \text{workint} + \text{rp\_heduc}$$
$$+ \text{rp\_occstat} + \text{rp\_branch} + \text{urb} + \text{migrbal} + \text{nonempm} \qquad (15)$$
$$+ \text{nonempw} + \text{quinc} + \text{carscc} + \text{unemp} + \text{compall}$$

The retransformed values are used again for the estimation of the at-risk-of-poverty rate for Austria and the nine regions.

(b) **Robust Linear Regression** (compare Chapter 2.3.2)

Now also some robust methods get considered. The response variable is again the equivalised household income EQ_INC.

(i) The first robust model corresponds to the first mentioned model with the socio-demographic variables rp_age, dwell, htyp, migration, equipm, rp_living, rp_famst, workint, rp_heduc, rp_occstat, rp_branch and urb as predictors. Again the same four variables are treated as ordered factors and the regression is calculated taking into account the weights. So the robust linear regression model for the income looks like in (14):

$$
\begin{aligned}
\log(\text{EQ\_INC}) \;\sim\;\; & \text{rp\_age} + \text{dwell} + \text{htyp} + \text{migration} + \text{equipm} \\
+\;\; & \text{rp\_living} + \text{rp\_famst} + \text{workint} + \text{rp\_heduc} \\
+\;\; & \text{rp\_occstat} + \text{rp\_branch} + \text{urb}
\end{aligned}
$$

Again the model gets estimated for the simulated EU-SILC data set and the calculated predictions for the simulated Micro Census data set are retransformed in the following with the exponential function. With these values once again the at-risk-of-poverty rate is estimated on the one hand for Austria and on the other hand also for every of the nine provinces. The difference to (a)(i) is in the estimation of the model, because in the actual case it is done using Huber's M-estimator instead of the ordinary least squares estimator.

(ii) As might be reasonably expected the next regarded model is the same as in (a)(ii), with the only difference, that the model is fitted by robust regression using an M-estimator, namely again Huber's M-estimator. Everthing else remains the same, so the model looks equally as in (15)

$$
\begin{aligned}
\log(\text{EQ\_INC}) \;\sim\;\; & \text{rp\_age} + \text{dwell} + \text{htyp} + \text{migration} + \text{equipm} \\
+\;\; & \text{rp\_living} + \text{rp\_famst} + \text{workint} + \text{rp\_heduc} \\
+\;\; & \text{rp\_occstat} + \text{rp\_branch} + \text{urb} + \text{migrbal} + \text{nonempm} \\
+\;\; & \text{nonempw} + \text{quinc} + \text{carscc} + \text{unemp} + \text{compall}
\end{aligned}
$$

With the same steps as above, the at-risk-of-poverty rate gets estimated for Austria and the nine states.

In general there exists far more methods concerning the robust linear regression, like the "MM"-estimators or the "S"-estimators, but these methods are not used, because the implementations of these methos lead to computational problems and crashes of the code.

(c) **Logistic Regression** (compare Chapter 2.3.3)

Henceforward the dependent variable is not the equivalised household income, but the binary variable ARPT60i. Analogous to the linear regression models two different models are regarded, the first one with only socio-demographic variables as predictors, the other one with additionally covariates.

(i) The first considered logistic regression model is similarly structured as the first OLS model (compare (a)(i)). Thus the variables rp_age, dwell, htyp, migration, equipm, rp_living, rp_famst, workint, rp_heduc, rp_occstat, rp_branch and urb are the predictor variables and the four factors rp_age, equipm, workint and rp_heduc are treated as ordered. But in contrast to the above mentioned model, the response is now the binary variable ARPT60i. So the model for the dichotomous variable results in:

$$
\begin{aligned}
\text{ARPT60i} \sim\ & \text{rp\_age} + \text{dwell} + \text{htyp} + \text{migration} + \text{equipm} \\
+\ & \text{rp\_living} + \text{rp\_famst} + \text{workint} + \text{rp\_heduc} \\
+\ & \text{rp\_occstat} + \text{rp\_branch} + \text{urb}
\end{aligned}
$$

Again weights are included in the model estimation and the model is calculated based on the simulated EU-SILC data set. As can be seen above, it gets transmitted to the artificial Micro Census data set. These imputed values of the dichotomous variable are the basis for the calculation of the at-risk-of-poverty rate for Austria and the nine provinces. The rates are calculated as weighted means.

(ii) As mentioned above, in context with logistic regression also a model with the just now used socio-demographic variables and some of the covariates explained in Chapter 4.1.3 is established. Again, not merely due to singularity problems and so on, but rather to provide comparability with the other methods and models, the same choice is included into the model, namely the covariates migrbal, nonempm, nonempw, quinc, carscc, unemp and compall. The same four factors as in the other so far considered models are treated as ordered and weights are incorporated in the model estimation. So the second model for the dichotomous variable ARPT60i as response results in:

$$
\begin{aligned}
\text{ARPT60i} \sim\ & \text{rp\_age} + \text{dwell} + \text{htyp} + \text{migration} + \text{equipm} \\
+\ & \text{rp\_living} + \text{rp\_famst} + \text{workint} + \text{rp\_heduc} \\
+\ & \text{rp\_occstat} + \text{rp\_branch} + \text{urb} + \text{migrbal} + \text{nonempm} \\
+\ & \text{nonempw} + \text{quinc} + \text{carscc} + \text{unemp} + \text{compall}
\end{aligned}
$$

In the same way as above the model is estimated using the simulated EU-SILC data set and then it is transmitted to the simulated Micro Census data set. Of some and all, respectively, of the predicted values the weighted mean is calculated to get the at-risk-of-poverty rate for the nine regions and Austria, respectively.

The following implemented models refer to mixed regression. On the one side again models with the equivalised household income as respone are regarded and on the other side models with the factor at-risk-of-poverty as dependent variable get considered. For both alternatives one version exists with the above described and already used socio-demographic variables and one version with additionally covariates.

(d) **Linear Mixed Regression** (compare Chapter 2.3.4)

Concerning the linear mixed models, the response is again the equivalised household income EQ_INC.

(i) The same factors as in the previous models are concerned as ordered and due to comparableness the same socio-demographic variables as before are used to build the first model in this vein. In mixed effects models the idea is to include some random effects to the fixed effects. So it is worked with a model where the socio-demographic variables are treated as the fixed effects and the variable strat, that belongs to the region of Austria, as the random effect with notation (1|strat), see below. Weights are included in the model estimation. The resulting model for linear mixed regression with EQ_INC as dependent variable calculated on the simulated EU-SILC data set, where of course the log-transformed values of the income are taken, looks like

$$
\begin{aligned}
\log(\text{EQ\_INC}) \sim\ & \text{rp\_age} + \text{dwell} + \text{htyp} + \text{migration} + \text{equipm} \\
+\ & \text{rp\_living} + \text{rp\_famst} + \text{workint} + \text{rp\_heduc} \\
+\ & \text{rp\_occstat} + \text{rp\_branch} + \text{urb} + (1|\text{strat})
\end{aligned}
$$

The obtained model is used for the imputation of the values of log(EQ_INC) into the Micro Census data set and these values get retransformed with the exponential function. Then the at-risk-of-poverty rate is estimated for Austria and the nine regions out of these values using the corresponding weights.

(ii) Following the above listed models, now a model is regarded, where some of the covariates are added to the model explained in (d)(i). So the independent variables rp_age, dwell, htyp, migration, equipm, rp_living, rp_famst, workint, rp_heduc, rp_occstat, rp_branch and urb in collaboration with the choice of covariates migrbal, nonempm, nonempw, quinc, carscc, unemp and compall are seen as the fixed effects where the factors rp_age, equipm, workint and rp_heduc are used as ordered again. As random effect again the variable strat is chosen. For the estimation of the model, using the artificial EU-SILC data set, weights are included once again. So now the model for log(EQ_INC) is written as

$$
\begin{aligned}
\log(\text{EQ\_INC}) \sim\ & \text{rp\_age} + \text{dwell} + \text{htyp} + \text{migration} + \text{equipm} \\
+\ & \text{rp\_living} + \text{rp\_famst} + \text{workint} + \text{rp\_heduc} \\
+\ & \text{rp\_occstat} + \text{rp\_branch} + \text{urb} + \text{migrbal} + \text{nonempm} \\
+\ & \text{nonempw} + \text{quinc} + \text{carscc} + \text{unemp} + \text{compall} + (1|\text{strat})
\end{aligned}
$$

This model gets transmitted to the simulated Micro Census data set and the obtained values are naturally retransformed. Afresh the at-risk-of-poverty rate of Austria and every of the nine provinces is estimated with these values incorporating the weights.

(e) **Logistic Mixed Regression** (compare Chapter 2.3.5)

Unlike linear mixed regression models regarding logistic mixed regression the response is no longer the equivalised household income, but the factor at-risk-of-poverty. Apart from that the implemented models are structured in the same way.

(i) Also for the last regarded regression method the factors rp_age, dwell, htyp, migration, equipm, rp_living, rp_famst, workint, rp_heduc, rp_occstat, rp_branch and urb are incorporated as the fixed effects in the model (whereby the same four factors are treated as ordered factors) and the variable strat as random effect. As a result the first logistic mixed regression model looks like

$$\text{ARPT60i} \sim \text{rp\_age} + \text{dwell} + \text{htyp} + \text{migration} + \text{equipm}$$
$$+ \text{rp\_living} + \text{rp\_famst} + \text{workint} + \text{rp\_heduc}$$
$$+ \text{rp\_occstat} + \text{rp\_branch} + \text{urb} + (1|\text{strat})$$

By analogy with the other models weights are used for the estimation of the model based on the EU-SILC data set. The model gets transmitted to the Micro Census data set and the values are used for the calculations of the at-risk-of-poverty rates for Austria and the nine states, which are done in form of a weighted means.

(ii) For the second model of the last regression method the covariates migrbal, nonempm, nonempw, quinc, carscc, unemp and compall are added as fixed effects in comparison to (e)(i). The rest - i.e. the factor at-risk-of-poverty as dependent variable, the other fixed effects (partially as ordered factors), the single random effect and the usage of weights - remains the same. So the model looks like

$$\text{ARPT60i} \sim \text{rp\_age} + \text{dwell} + \text{htyp} + \text{migration} + \text{equipm}$$
$$+ \text{rp\_living} + \text{rp\_famst} + \text{workint} + \text{rp\_heduc}$$
$$+ \text{rp\_occstat} + \text{rp\_branch} + \text{urb} + \text{migrbal} + \text{nonempm}$$
$$+ \text{nonempw} + \text{quinc} + \text{carscc} + \text{unemp} + \text{compall} + (1|\text{strat})$$

Again it is estimated on the EU-SILC data set and gets transmitted to the artificial Micro Census data set. The imputed values of the dichotomous variable are the basis for the calculation of the at-risk-of-poverty rate for Austria and the nine provinces. These rates are calculated once again as weighted means.

4.2.2. Statistical Matching

As already mentioned and described above, due to the fact, that the aim is to estimate the at-risk-of-poverty rate, which is calculated with the equivalised household income or with the binary variable at-risk-of-poverty, there are two possibilities:

1. Values for the equivalised household income get imputed in the simulated Micro Census data sets and then the at-risk-of-poverty rate is estimated.

2. First the variable at-risk-of-poverty is calculated for the observations in the artificial EU-SILC data sets and these values are taken as basis for the imputation of at-risk-of-poverty in the Micro Census data set. Then out of this dichotomous variable the at-risk-of-poverty rate gets calculated.

Because of the large number of available variables in the data sets and the variety of different methods, there are a lot of possibilities to perform the imputation. Several approaches get considered using the above described variables (in the Chapter 4.1.3). So artificial Micro Census und EU-SILC data sets with the variables rp_age, dwell, htyp, migration, equipm, rp_living, rp_famst, workint, rp_heduc, rp_occstat, rp_branch, urb, the covariates migrbal, nonempm, nonempw, birth, mort, quinc, carscc, unemp, socass, compall, unemp_lag, socass_lag, compall_lag, the state state and the weight gew are considered. The EU-SILC data set includes furthermore the variables EQ_INC and ARPT60i, that are used for the calculation of the at-risk-of-poverty rate.

(f) Random Hot Deck

The first considered method is random hot deck using diverse imputation variables and donation classes.

(i) First of all the factor ARPT60i is imputed in the artificial Micro Census data set with the EU-SILC data set as donor file. For that reason the variable strat, i.e. the state of Austria, is used for building domains and it get imputed within these groups. The imputed values are the basis for the computation of the at-risk-of-poverty rates (in form of weighted means) for Austria and every of the nine states.

(ii) For the second model the same donation class strat is used, but now values for the numeric variable EQ_INC are imputed in the artificial Micro Census data set. With these values the at-risk-of-poverty rates for Austria and the nine provinces are estimated incorporating the weights.

(iii) The next considered model is the same as in (i) with the difference that strat and migration, i.e. the foreign origin of the household (divided in 3 levels: Austria, rest of EU and non-EU), are used as variables for building domains. So for the variable ARPT60i the missing values should be imputed and after that they are used for the calculation of the at-risk-of-poverty rates.

(iv) As may be assumed the following imputation concerns the variable EQ_INC with the variables strat and migration used for the formation of donation classes. Again out of the imputed values with the assistance of weights the at-risk-of-poverty rates for Austria and the states are estimated.

(g) Sequential Random Hot Deck

Furthermore 4 versions using the sequential random hot deck algorithm are considered. In distinction from the models considered in (f) "Random Hot Deck" additional ordering variables are defined.

(i) The first version is arranged as follows: The variable where missing values should be imputed is the factor at-risk-of-poverty and the used donation classes are configured with the variables strat and migration. As ordering variables rp_occstat, workint, rp_heduc, rp_living, rp_age and equipm are chosen. Just as a quick reminder, these variables are the occupational status, the work intensity of the household, the highest completed level of education, the living, the age class and the category of the equipment of the apartment. With the imputed values as basis the at-risk-of-poverty rates get calculated as weighted means for Austria and the nine states seperately.

(ii) The second version is distinct from the last regarded model (i.e. the first version (i) in (g) "Sequential Random Hot Deck") only to the effect that instead of the dichotomous factor ARPT60i the variable equivalised household income EQ_INC is imputed and these values are used for the calculation of the rates for Austria and the states.

(iii) Next again the factor at-risk-of-poverty gets imputed in the simulated Micro Census data set. The domain groups remain the same. The used ordering variables are nearly all in the simulated data sets occuring variables, namely rp_age, dwell, htyp, migration, equipm, rp_living, rp_famst, workint, rp_heduc, rp_occstat, rp_branch, migrbal, nonempm, nonempw, birth, mort, quinc, carscc, unemp, socass, compall, unemp_lag, socass_lag, compall_lag and strat. So the result of the imputation with these conditions is used for the calculation of the weighted means, i.e. of the at-risk-of-poverty rates.

(iv) The last considered version related to sequential random hot deck imputes values for the numeric variable EQ_INC. Here once again strat and migration play the role of the variables for building domains and impute within these domains. As ordering variables the same 25 as in the most recently version are used. Out of the imputed values the at-risk-of-poverty rates for Austria and the nine states are calculated.

(h) Weighted Random Hot Deck

The last applied method in the context of estimating the at-risk-of-poverty rate for Austria and the nine states is the weighted random hot deck method. As distinguished from the "normal" random hot deck, the choice is no longer completely at random, but the weights are used to pick a donor with corresponding proportional propabilities.

(i) In the first version of weighted random hot deck models the factor at-risk-of-poverty gets imputed. The variable `strat`, i.e. the state of Austria, is employed to identify donation classes. There is no use of matching variables, so all the units in the same donation class are possible donors. As a result one of them is selected with probability proportional to its weight. After the imputation of the values in the artificial Micro Census data set the at-risk-of-poverty rate gets estimated in form of a weighted mean for Austria as well as for every of the nine states.

(ii) The next version distinguishes from (i) only in the variable that should be "donated", so the variable that should be imputed. Here it is defined as the numeric variable `EQ_INC`, the equivalised household income. Hence again the states mould the donation classes, no matching variables get incorporated and the weights provide the probabilities of the donors to be chosen. The imputed values are the basis for the calculation of the at-risk-of-poverty threshold, the factor at-risk-of-poverty and the at-risk-of-poverty rate for Austria and the states.

(iii) Now again the factor at-risk-of-poverty gets imputed with the method of weighted random hot deck. Again no matching variables are used and hence all observations of one domain are considered as possible donors. In this version domains are built with the factor variable `strat`, i.e. the state of Austria, and the factor variable `migration`, i.e. the foreign origin of the household. Weighted means of the imputed at-risk-of-poverty values result in the estimations of the at-risk-of-poverty rates for Austria and the nine provinces.

(iv) As might be reasonably expected the last used method looks as follows: The numeric variable `EQ_INC`, i.e. the equivalised household income, is imputed again, the variables `strat` and `migration` are used to define donation classes, no matching variables are used and as a result a donor is selected with probability proportional to its weight. The at-risk-of-poverty rates for Austria and the nine states are computed out of the imputed values.

Due to the high computation time, no further models are employed in the context of weighted random hot deck.

(i) Constrained methods

Due to the fact, that the donor file (equates to the EU-SILC data set) is smaller than the record file (equates to the Micro Census data set), it is not possible to use constrained methods. Naturally one could not impute every value maximal one time, if there are more missing values than available values.

4.3. Results

In this chapter the results of the several methods get evaluated and compared in terms of bias and variance (see Chapter 3.3).

The aim is to find the "best" method in the sense of small bias and small variance.

Furthermore the reliability of the bootstrapping is checked for all methods.

The corresponding R-Code can be found in the Appendix A.1. The different models are explained in Appendix A.1.1 and A.1.2, the additional R-Code and the computation of the quality criteria is in Appendix A.1.3.

4.3.1. Reliability of the Bootstrapping

For the check of the reliability of the bootstrapping, the difference in the mean of the standard deviations yielded from the bootstrap replicates and the standard deviation estimated on the basis of the results of the J drawn samples is computed, see also Equation 8 and Equation 9. It should be nearly 0. Furthermore, the differences in the mean of the at-risk-of-poverty rates from the bootstrap replicates (of all drawn samples) and the mean of the at-risk-of-poverty rates estimated on the basis of the results of the repeated sample drawing are computed and evaluated, see also Equation 10 and Equation 11.

	Austria	Burgenland	Lower Austria	Vienna	Carinthia
DSD_S (Eq. 9)	0.03851	-0.03389	0.04235	0.04125	0.10708

	Styria	Upper Austria	Salzburg	Tyrol	Vorarlberg
DSD_S (Eq. 9)	0.03112	-0.00669	0.04332	0.11300	0.09736

Table 2: Difference in the mean of the standard deviations yielded from the bootstrap replicates and the standard deviation calculated on the basis of the results of the repeated EU-SILC sample drawing [in %].

As can be seen in the resulting figures in Table 2, Table 3 and Table 4, the bootstrapping regarding the variance works reliably. For Austria the absolute value of the difference concerning the standard deviation ranges from 0.01330535% (-0.01330535% for the robust linear regression model (b)(i)) to 0.4412739% (logistic regression model (c)(i)).

For the several states similar figures are obtained, see Table 2, Table 3 and Table 4.

	Austria	Burgenland	Lower Aus.	Vienna	Carinthia
(a)(i) Linear Regression	0.05574	0.12990	0.05123	0.03500	0.13485
(a)(ii) Linear Regression	0.05025	0.09615	0.00903	0.04285	0.21817
(b)(i) Robust Lin.Reg.	-0.01331	0.01891	-0.02310	-0.04582	0.01644
(b)(ii) Robust Lin.Reg.	-0.01405	-0.16265	-0.05007	-0.03349	-0.06676
(c)(i) Logistic Reg.	0.44127	0.53477	0.48001	0.52845	0.47652
(c)(ii) Logistic Reg.	0.33996	0.29944	0.30813	0.42043	0.55253
(d)(i) Linear Mixed Reg.	0.05608	0.13458	0.05565	0.05798	0.13323
(d)(ii) Linear Mixed Reg.	0.05025	0.09615	0.00903	0.04285	0.21817
(e)(i) Logistic Mix.Reg.	0.40088	0.57513	0.40499	0.50224	0.45202
(e)(ii) Logistic Mix.Reg.	0.38181	0.38591	0.24537	0.48611	0.56691
(f)(i) Random Hot Deck	0.10759	0.35070	0.18465	0.08930	0.33142
(f)(ii) Random Hot Deck	0.09021	0.07959	0.19268	0.18025	0.27532
(f)(iii) Random Hot Deck	0.07879	0.06130	0.15268	0.20184	0.43642
(f)(iv) Random Hot Deck	0.08058	0.29505	0.20936	0.27124	0.12309
(g)(i) Sequential R.H.D.	-0.03280	-0.30341	-0.30457	-0.34157	-0.14682
(g)(ii) Sequential R.H.D.	-0.06330	-0.31657	-0.25072	-0.32082	-0.14923
(g)(iii) Sequential R.H.D.	0.05341	-0.16091	-0.01933	0.04860	0.00408
(g)(iv) Sequential R.H.D.	-0.01632	-0.11299	0.00315	0.09634	-0.00250
(h)(i) Weighted R.H.D.	0.06206	0.09661	0.09740	0.15023	0.15009
(h)(ii) Weighted R.H.D.	0.08935	0.05727	0.22385	0.21695	0.23609
(h)(iii) Weighted R.H.D.	0.06245	0.12738	0.12139	0.18943	0.20686
(h)(iv) Weighted R.H.D.	0.09098	0.14992	0.17686	0.19363	0.31405

Table 3: Difference in the mean of the standard deviations yielded from the bootstrap replicates and the standard deviation calculated on the basis of the results of the repeated Micro Census sample drawing [in %] - part 1: Austria, Burgenland, Lower Austria, Vienna, Carinthia.

In most of the cases the robust linear regression models (b)(i) and (b)(ii) have the minimal absolute difference. Only for Lower Austria and Carinthia the sequential random hot deck model (g)(iv) and for Styria the linear regression model (a)(ii) result in the minimal absolute difference.

The maximal absolute difference is attained by the logistic regression model (c)(i) for 7 states. Furthermore for any state the maximum is obtained once with the logistic mixed regression model (e)(i) and also once with the logistic mixed regression model (e)(ii).

Looking at the average difference of the nine states for every model (see Appendix A.2, Table 30), the following ascending order of the absolute values results: robust linear regression model (b)(i), sequential random hot deck model (g)(iv), sequential random hot deck model (g)(iii), robust linear regression model (b)(ii), linear mixed regression model (d)(ii), linear regression model (a)(ii), linear regression model (a)(i), linear mixed regression model (d)(i), weighted random hot deck model (h)(i), weighted random hot deck model (h)(iii), random hot deck model (f)(iii), weighted random hot deck model

	Styria	Upper Aus.	Salzburg	Tyrol	Vorarlberg
(a)(i) Linear Regression	0.11172	0.10002	0.09942	0.11096	0.14065
(a)(ii) Linear Regression	0.00312	0.03206	0.07465	0.12355	0.03817
(b)(i) Robust Lin.Reg.	0.03433	0.00323	0.02067	0.02352	0.02652
(b)(ii) Robust Lin.Reg.	-0.01054	-0.04672	-0.05828	-0.01395	-0.16344
(c)(i) Logistic Reg.	0.47334	0.48715	0.49260	0.54641	0.59075
(c)(ii) Logistic Reg.	0.25917	0.23568	0.30706	0.45995	0.39907
(d)(i) Linear Mixed Reg.	0.11294	0.10029	0.10247	0.10475	0.14594
(d)(ii) Linear Mixed Reg.	0.00312	0.03206	0.07465	0.12355	0.03817
(e)(i) Logistic Mix.Reg.	0.40316	0.43480	0.48941	0.51678	0.53866
(e)(ii) Logistic Mix.Reg.	0.38369	0.27552	0.27349	0.49271	0.47551
(f)(i) Random Hot Deck	0.21074	0.03128	0.24224	0.39059	0.11930
(f)(ii) Random Hot Deck	0.25550	0.19996	0.25417	0.33911	0.11208
(f)(iii) Random Hot Deck	0.18976	0.08615	0.15332	0.19226	0.12113
(f)(iv) Random Hot Deck	0.03009	0.11951	0.23191	0.27614	0.21715
(g)(i) Sequential R.H.D.	-0.37705	-0.28186	-0.26749	-0.31527	-0.24625
(g)(ii) Sequential R.H.D.	-0.27716	-0.21371	-0.20734	-0.29053	-0.25798
(g)(iii) Sequential R.H.D.	-0.09930	-0.06890	-0.04540	-0.08076	0.05448
(g)(iv) Sequential R.H.D.	-0.01533	-0.05296	0.03698	-0.05079	-0.04208
(h)(i) Weighted R.H.D.	0.15378	0.07067	0.13458	0.15680	0.17799
(h)(ii) Weighted R.H.D.	0.18495	0.09668	0.22891	0.23099	0.15175
(h)(iii) Weighted R.H.D.	0.12874	0.16629	0.11398	0.11500	0.20837
(h)(iv) Weighted R.H.D.	0.18137	0.15932	0.20786	0.29199	0.19385

Table 4: Difference in the mean of the standard deviations yielded from the bootstrap replicates and the standard deviation calculated on the basis of the results of the repeated Micro Census sample drawing [in %] - part 2: Styria, Upper Austria, Salzburg, Tyrol, Vorarlberg.

	Austria	Burgenland	Lower Austria	Vienna	Carinthia
DM_S (Eq. 11)	0.051859	0.102556	0.036913	0.054772	0.083648

	Styria	Upper Austria	Salzburg	Tyrol	Vorarlberg
DM_S (Eq. 11)	0.044448	0.040111	0.035370	0.060470	0.093183

Table 5: Difference in the mean of the at-risk-of-poverty rates from the bootstrap replicates (of all drawn samples) and the mean of the at-risk-of-poverty rates calculated on the basis of the results of the repeated EU-SILC sample drawing [in %].

(h)(ii), random hot deck model (f)(iv), weighted random hot deck model (h)(iv), random hot deck model (f)(ii), random hot deck model (f)(i), sequential random hot deck model (g)(ii), sequential random hot deck model (g)(i), logistic regression model (c)(ii), logistic mixed regression model (e)(ii), logistic mixed regression model (e)(i), logistic regression model (c)(i).

Now the difference in the mean of the at-risk-of-poverty rates yielded from the bootstrap replicates (of all drawn samples) and the mean of the at-risk-of-poverty rates calculated on the basis of the results of the repeated sample drawing are considered, see also Equation 10 and Equation 11.

For Austria the absolute value of this difference ranges from 0.02049814% (logistic mixed regression model (e)(i)) to 0.56678441% (−0.56678441% for the robust linear regression model (b)(i)), see Table 5 and Table 6. For the direct estimator the difference is 0.05185870%, see Table 5.

	Austria	Burgenland	Lower Aus.	Vienna	Carinthia
(a)(i) Linear Regression	0.26395	0.31577	0.28685	0.23596	0.29775
(a)(ii) Linear Regression	0.28993	0.44871	0.28174	0.22534	0.43336
(b)(i) Robust Lin.Reg.	-0.56678	-0.43937	-0.40755	-0.87697	-0.41559
(b)(ii) Robust Lin.Reg.	-0.56283	-0.21234	-0.43276	-0.90465	-0.43530
(c)(i) Logistic Reg.	0.05659	0.13478	0.07019	0.05562	0.06660
(c)(ii) Logistic Reg.	0.04910	0.18107	0.06709	0.03760	0.07548
(d)(i) Linear Mixed Reg.	0.25391	0.31020	0.28485	0.18629	0.30894
(d)(ii) Linear Mixed Reg.	0.28993	0.44871	0.28174	0.22534	0.43336
(e)(i) Logistic Mix.Reg.	0.02050	-0.05069	0.01035	0.04193	-0.04296
(e)(ii) Logistic Mix.Reg.	0.05372	0.14271	0.03000	0.08492	0.04863
(f)(i) Random Hot Deck	0.07591	0.04709	0.02906	0.05034	0.06165
(f)(ii) Random Hot Deck	0.06245	0.41971	0.00350	0.05257	0.05880
(f)(iii) Random Hot Deck	0.09359	0.11748	-0.01468	-0.05792	0.32790
(f)(iv) Random Hot Deck	0.06180	0.05835	0.07464	0.05382	0.26390
(g)(i) Sequential R.H.D.	-0.27523	-0.16305	-0.44558	-0.17605	-0.61723
(g)(ii) Sequential R.H.D.	-0.19993	-0.04772	-0.40149	-0.02482	-0.48056
(g)(iii) Sequential R.H.D.	0.02071	0.03945	-0.01087	-0.02060	0.07587
(g)(iv) Sequential R.H.D.	0.03166	0.02343	0.00789	0.02267	0.09755
(h)(i) Weighted R.H.D.	0.06571	0.08417	0.04203	0.03412	0.06039
(h)(ii) Weighted R.H.D.	0.05825	-0.04667	0.06149	0.08569	0.03203
(h)(iii) Weighted R.H.D.	0.02663	0.04784	0.01508	0.06876	0.00888
(h)(iv) Weighted R.H.D.	0.08320	0.09992	0.09827	0.08854	0.07930

Table 6: Difference in the mean of the at-risk-of-poverty rates from the bootstrap replicates (of all drawn samples) and the mean of the at-risk-of-poverty rates calculated on the basis of the results of the repeated Micro Census sample drawing [in %]- part 1: Austria, Burgenland, Lower Austria, Vienna, Carinthia.

The absolute value of the difference for any state or Austria ranges from 1.203364e-07% (−1.203364e-07% is obtained by the logistic regression model (c)(i) for Vorarlberg) to 0.9046489% (−0.9046489% is obtained by the robust linear regression model (b)(ii) for Vienna), see Table 5, Table 6 and Table 7.

	Styria	Upper Aus.	Salzburg	Tyrol	Vorarlberg
(a)(i) Linear Regression	0.24156	0.22372	0.30429	0.31102	0.28487
(a)(ii) Linear Regression	0.24158	0.23935	0.33375	0.35035	0.46133
(b)(i) Robust Lin.Reg.	-0.45358	-0.60929	-0.46923	-0.45393	-0.70151
(b)(ii) Robust Lin.Reg.	-0.46321	-0.64427	-0.46608	-0.30444	-0.63876
(c)(i) Logistic Reg.	0.04783	0.02494	0.12334	0.04750	-1.203e-07
(c)(ii) Logistic Reg.	-0.01032	0.05070	0.13434	5.155e-04	0.03878
(d)(i) Linear Mixed Reg.	0.25022	0.22857	0.29655	0.30192	0.26476
(d)(ii) Linear Mixed Reg.	0.24158	0.23935	0.33375	0.35035	0.46133
(e)(i) Logistic Mix.Reg.	0.02899	0.04568	0.04138	-0.00534	0.00980
(e)(ii) Logistic Mix.Reg.	-0.00227	0.02437	0.00350	0.14272	0.14942
(f)(i) Random Hot Deck	0.15132	-0.00557	0.08937	0.18069	0.08870
(f)(ii) Random Hot Deck	0.20775	-0.03060	0.19824	-0.03240	-0.02441
(f)(iii) Random Hot Deck	0.08956	0.02350	0.07702	0.06298	0.29897
(f)(iv) Random Hot Deck	-0.02124	0.04532	0.01960	0.08294	0.09147
(g)(i) Sequential R.H.D.	-0.26049	-0.41729	-0.23185	-0.05709	-0.06525
(g)(ii) Sequential R.H.D.	-0.17880	-0.33420	-0.14213	0.06380	0.04525
(g)(iii) Sequential R.H.D.	-0.01805	0.05462	-0.02229	0.02516	0.08354
(g)(iv) Sequential R.H.D.	0.00694	0.05943	-0.00349	0.02887	0.11047
(h)(i) Weighted R.H.D.	0.05238	0.08000	0.06604	0.03540	0.15897
(h)(ii) Weighted R.H.D.	0.08276	0.04792	0.02253	0.06592	0.03480
(h)(iii) Weighted R.H.D.	0.01226	0.05752	0.02756	-0.05032	0.05090
(h)(iv) Weighted R.H.D.	0.05636	0.07725	0.06373	0.10355	0.08487

Table 7: Difference in the mean of the at-risk-of-poverty rates from the bootstrap replicates (of all drawn samples) and the mean of the at-risk-of-poverty rates calculated on the basis of the results of the repeated Micro Census sample drawing [in %] - part 2: Styria, Upper Austria, Salzburg, Tyrol, Vorarlberg.

Table 8 shows for every state the models that take the minimal and maximal absolute value of the difference in the means of the at-risk-of-poverty rates.

Detailed analysis of the figures and accordingly computing the mean of the several models for the nine states (see also Appendix A.2, Table 31) show that the absolute difference is greater for smaller states, like for example Burgenland, Vorarlberg or Tyrol, than for the bigger states, like Lower Austria, Styria, Vienna or Upper Austria. Furthermore the attention is attracted by the sign of this mean: the greater states (Upper Austria, Vienna and Lower Austria) have a negative mean of the differences of the models.

All in all the bootstrap-error is very small and it decreases by increasing the number of samples and the number of bootstrap replicates. It can be concluded that no bias is introduced by the bootstrap used in the simulations.

	minimum	maximum
Austria	(e)(i) Logistic Mixed Regr.	(b)(i) Robust Lin. Regr.
Burgenland	(g)(iv) Sequential R. H. D.	(a)(ii) Linear Regression
Lower Austria	(f)(ii) Random Hot Deck	(g)(i) Sequential R. H. D.
Vienna	(g)(iii) Sequential R. H. D.	(b)(ii) Robust Lin. Regr.
Carinthia	(h)(iii) Weighted R. H. D.	(g)(i) Sequential R. H. D.
Styria	(e)(ii) Logistic Mixed Regr.	(b)(ii) Robust Lin. Regr.
Upper Austria	(f)(i) Random Hot Deck	(b)(ii) Robust Lin. Regr.
Salzburg	(g)(iv) Sequential R. H. D.	(b)(i) Robust Lin. Regr.
Tyrol	(c)(ii) Logistic Regr.	(b)(i) Robust Lin. Regr.
Vorarlberg	(c)(i) Logistic Regr.	(b)(i) Robust Lin. Regr.

Table 8: Models with the minimum / maximum absolute value of the difference in the mean of the at-risk-of-poverty rates from the bootstrap replicates (of all drawn samples) and the mean of the at-risk-of-poverty rates calculated on the basis of the results of the repeated sample drawing for Austria and the nine states.

4.3.2. Bias

For the computation of the bias, the "true" at-risk-of-poverty rate is calculated first based on the close-to-reality simulated population U, once for Austria and also for every of the 9 states separately, see Listing 25 in Appendix A.1.3 and Table 9.

	Austria	Burgenland	Lower Austria	Vienna	Carinthia
arprPOP	12.563	16.447	9.626	16.523	17.434

	Styria	Upper Austria	Salzburg	Tyrol	Vorarlberg
arprPOP	11.040	8.930	11.145	14.395	14.223

Table 9: At-risk-of-poverty rates of the artificial population [in %].

Furthermore, the bias gets estimated on the basis of the EU-SILC samples and also on basis of the estimated values of the Micro Census data sets (see also Equation 3 and 4). Thus, two approaches are considered.

(1) The bias is estimated by the mean of the at-risk-at-poverty rates obtained from the repeated sampling of EU-SILC minus the "true" at-risk-at-poverty rate (obtained from U). This is done for Austria and also for each state.

(2) The same as in (1) but using the (equivalized income extended) Micro Census data. See also Listing 24 and Listing 25 in Appendix A.1.3.

Looking at the bias obtained from the EU-SILC samples someone notices that it ranges between 0.03731879% and 0.30632906% (compare Table 10). It should be nearly 0, because it is based on the direct extrapolation.

	Austria	Burgenland	Lower Austria	Vienna	Carinthia
BiasS	0.12682	0.12085	0.07772	0.28307	0.03732

	Styria	Upper Austria	Salzburg	Tyrol	Vorarlberg
BiasS	0.10101	0.03782	0.08581	0.09420	0.30633

Table 10: Bias on basis of the EU-SILC samples [in %].

Comparing the bias obtained from the different methods for the Micro Census samples with the bias obtained from the EU-SILC samples it can be noticed that some of the methods do well, better than the direct estimation, and some other methods don't perform as good as the direct estimation using EU-SILC samples, see Tables 11 and 12.

Regression Models

The linear regression models (a)(i) and (a)(ii) as well as the robust linear regression models (b)(i) and (b)(ii) result in a larger absolute value of the bias for Austria (see Table 11) than direct estimation (in Table 10). This higher bias is due to the fact that the equivalised household income cannot be predicted satisfactorily and therefore the at-risk-of-poverty threshold and rates (estimated on basis of these predicted values) are biased.

The logistic regression models (c)(i) and (c)(ii) perform much better, the absolute value of the bias for Austria (see Table 11) is even lower than the absolute value of the bias for Austria received from the EU-SILC samples (see Table 10). The binary variable at-risk-of-poverty can be predicted better.

The same pattern is found regarding the mixed regression models: By comparison the linear mixed regression models (d)(i) and (d)(ii) do worse and the logistic mixed regression models (e)(i) and (e)(ii) perform better for whole Austria, although model (e)(i) doesn't better than the direct estimation.

An interesting fact is that the linear regression models and the linear mixed regression models underestimate the "true" at-risk-of-poverty rate of Austria, as can be recognised by the minus-sign. The other regression models overestimate the rate, because the bias is positive, see Table 11 and Table 12. Similar observations are made looking at the several states, in the majority of cases the bias for the linear (mixed) regression models are negative and the bias for the logistic (mixed) regression models are positive. The underestimations result from the shift of the thresholds used for the estimations of the at-risk-of-poverty rates.

It turns out that in terms of the bias of Austria the logistic mixed regression model (e)(ii) performs best out of the investigated regression methods. This model results in a smaller absolute value of the bias than the direct estimator (based on the EU-SILC samples). Looking at the several states it is noticed that the absolute value of the bias of model (e)(ii) is also the smallest absolute value of the bias among the different regression

	Austria	Burgenland	Lower Aus.	Vienna	Carinthia
(a)(i) Linear Regression	-2.51227	-9.35414	-1.57561	-0.90658	-8.53200
(a)(ii) Linear Regression	-2.39992	-10.38204	-1.45135	-0.79622	-5.18893
(b)(i) Robust Lin.Reg.	-4.53377	-11.11499	-3.20755	-3.94370	-10.36583
(b)(ii) Robust Lin.Reg.	-4.45744	-11.96826	-3.09081	-3.85080	-9.50906
(c)(i) Logistic Reg.	0.11519	-5.64719	1.49070	0.22229	-5.34074
(c)(ii) Logistic Reg.	0.12005	0.09956	0.02278	0.23882	0.06426
(d)(i) Linear Mixed Reg.	-2.43169	-9.33726	-1.56030	-0.62823	-8.56516
(d)(ii) Linear Mixed Reg.	-2.39992	-10.38204	-1.45135	-0.79622	-5.18893
(e)(i) Logistic Mix.Reg.	0.80373	-4.78267	2.34431	0.21658	-4.50181
(e)(ii) Logistic Mix.Reg.	0.11349	0.14370	0.05289	0.19400	0.08196
(f)(i) Random Hot Deck	0.73664	0.17357	0.08154	0.28538	0.07408
(f)(ii) Random Hot Deck	0.17477	-0.08806	0.14203	0.35558	0.13251
(f)(iii) Random Hot Deck	0.73301	0.12423	0.13846	0.40459	-0.19045
(f)(iv) Random Hot Deck	0.17297	0.23459	0.08352	0.35535	-0.06897
(g)(i) Sequential R.H.D.	2.08358	0.82601	1.94299	1.59166	2.23548
(g)(ii) Sequential R.H.D.	0.74147	-0.02920	1.45624	0.56233	1.27195
(g)(iii) Sequential R.H.D.	0.69910	0.07207	0.18890	0.47653	-0.27892
(g)(iv) Sequential R.H.D.	0.20893	0.24757	0.24397	0.58107	-0.12740
(h)(i) Weighted R.H.D.	0.75486	0.15279	0.07877	0.29924	0.06932
(h)(ii) Weighted R.H.D.	0.14752	0.30579	0.07085	0.28596	0.12597
(h)(iii) Weighted R.H.D.	0.79221	0.16945	0.11046	0.26850	0.09980
(h)(iv) Weighted R.H.D.	0.12011	0.14262	0.03823	0.28818	0.05995

Table 11: Bias for the several models calculated on basis of the Micro Census data sets [in %] - part 1: Austria, Burgenland, Lower Austria, Vienna, Carinthia.

models for the states Vienna and Tyrol. The model estimation results again in a smaller absolute value of the bias than the direct estimation. The logistic regression model (c)(ii) performs best for the states Burgenland, Lower Austria, Carinthia, Upper Austria and Salzburg, the logistic mixed regression model (e)(i) does best for Vorarlberg and the logistic regression model (c)(i) performs best for Styria. The best model estimator for Burgenland, Lower Austria, Salzburg and Vorarlberg has a smaller absolute value of the bias than the bias received from the EU-SILC samples.

A further possibility of comparing the regression models is to evaluate the average of the bias of the nine states for every method (see also Appendix A.2, Table 32). It appears that now the logistic mixed regression model (e)(ii) performs best among the regression models, followed by the logistic regression model (c)(ii), the logistic mixed regression model (e)(i) and then the logistic regression model (c)(i).

Taking the absolute values for the computation of the average the result changes only a little bit (see also Appendix A.2, Table 32): Now again the logistic mixed regression model (e)(ii) performs best, followed by the logistic regression model (c)(ii), then by the logistic regression model (c)(i) and the logistic mixed regression model (e)(i).

44

After careful analysis it seems as if there is a "regression to the middle", so that the states with a higher at-risk-of-poverty rate, as for example Vienna or Carinthia, tend to get a lower rate and the states with a lower at-risk-of-poverty rate, as for example Lower Austria or Upper Austria, tend to get a higher rate. Hence the differences between the several states get underestimated.

Statistical Matching Methods

Comparing the bias for Austria among the statistical matching models, the weighted random hot deck model (h)(iv) has the smallest absolute value (see Table 11). Moreover, the absolute value of the bias is smaller as the one from the direct estimator of the at-risk-of-poverty rate for Austria. The other statistical matching models result in a higher absolute value of the bias (for Austria) than the one received from the EU-SILC samples.

	Styria	Upper Aus.	Salzburg	Tyrol	Vorarlberg
(a)(i) Linear Regression	-2.92145	0.23570	-2.61513	-5.38656	-3.43526
(a)(ii) Linear Regression	-1.02419	-0.60208	-3.98952	-5.69513	-6.52459
(b)(i) Robust Lin.Reg.	-4.80327	-1.49916	-4.18729	-6.92337	-6.05512
(b)(ii) Robust Lin.Reg.	-2.83822	-2.28936	-5.38743	-6.92675	-7.44453
(c)(i) Logistic Reg.	0.11999	2.74800	0.91722	-2.04878	-0.97900
(c)(ii) Logistic Reg.	0.14773	0.05903	-0.02546	0.17203	0.35063
(d)(i) Linear Mixed Reg.	-2.94049	0.28383	-2.54029	-5.32041	-3.28325
(d)(ii) Linear Mixed Reg.	-1.02419	-0.60208	-3.98952	-5.69513	-6.52459
(e)(i) Logistic Mix.Reg.	0.94116	3.65465	1.86695	-1.24110	-0.02816
(e)(ii) Logistic Mix.Reg.	0.13349	0.08535	0.11394	0.01901	0.25238
(f)(i) Random Hot Deck	-8.753e-04	0.07551	0.02189	-0.04276	0.28806
(f)(ii) Random Hot Deck	0.00482	0.16245	-0.02590	0.23380	0.51546
(f)(iii) Random Hot Deck	0.05416	0.06195	0.07321	0.08614	0.10253
(f)(iv) Random Hot Deck	0.23951	0.08349	0.13761	0.08882	0.37998
(g)(i) Sequential R.H.D.	1.64611	1.28486	1.18498	0.87267	1.25327
(g)(ii) Sequential R.H.D.	0.87188	0.57497	0.50183	-0.02499	0.27323
(g)(iii) Sequential R.H.D.	0.05266	-0.08960	-0.06089	-0.10340	0.27818
(g)(iv) Sequential R.H.D.	0.14908	0.01024	0.01160	0.00815	0.44054
(h)(i) Weighted R.H.D.	0.09795	-0.00558	0.06424	0.11421	0.24342
(h)(ii) Weighted R.H.D.	0.09030	0.04852	0.11952	0.12013	0.41242
(h)(iii) Weighted R.H.D.	0.13184	0.01768	0.10511	0.19417	0.36917
(h)(iv) Weighted R.H.D.	0.11354	0.01742	0.08218	0.06160	0.35712

Table 12: Bias for the several models calculated on basis of the Micro Census data sets [in %] - part 2: Styria, Upper Austria, Salzburg, Tyrol, Vorarlberg.

Looking at the several states it is noticed that as best model of the statistical matching methods concerning the bias of Burgenland emerges the sequential random hot deck model (g)(ii). It has a respectively smaller absolute value of the bias than the direct estimator. The absolute value of the bias for Salzburg and Tyrol is smallest for the estimations based on the sequential random hot deck model (g)(iv) and the respective absolute value of the bias is again smaller than the one received from the EU-SILC samples. The best statistical matching estimator in terms of the bias for Vienna is the weighted random hot deck model (h)(iii), in terms of the bias for Styria it is the random hot deck model (f)(i), in terms of the bias for Upper Austria it is the weighted random hot deck model (h)(i) and in terms of the bias for Vorarlberg the random hot deck model (f)(iii) is best. Once again the models result in a particular smaller absolute value of the bias than the direct estimator. For Lower Austria and Carinthia the weighted random hot deck model (h)(iv) performs best. Here can be found the sole exception regarding the comparison of the best statistical matching estimators for the different states with the direct estimator: The absolute value of the bias of the best estimator for Carinthia is higher than the absolute value of the bias received from the EU-SILC samples, but it is necessary to notice that the comparison concerns the fourth decimal place.

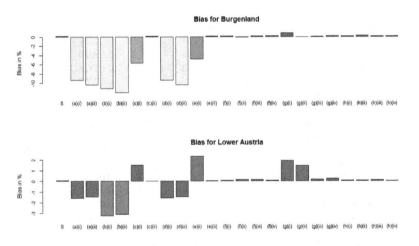

Figure 4: Bar chart of the bias [in %] for the direct estimator (calculated on basis of the EU-SILC data sets)(denoted by S) and the several models (calculated on basis of the Micro Census data sets) using the example of Burgenland and Lower Austria.

Looking again at the average of the nine states (see also Appendix A.2, Table 32) the sequential random hot deck model (g)(iii) performs best, followed by the random hot deck model (f)(iii), the random hot deck model (f)(i) and then by the weighted random

46

hot deck models (h)(i) and (h)(iv).

Taking the absolute values the result changes again (see also Appendix A.2, Table 32): the random hot deck model (f)(i) does best, followed by the weighted random hot deck models (h)(i) and (h)(iv), then by the random hot deck model (f)(iii), the weighted random hot deck models (h)(iii), (h)(ii) and the sequential random hot deck model (g)(iii).

Comparison: Regression Models and Statistical Matching Methods

It is striking that in general there is not such a great difference between the (absolute values of the) bias for the several statistical matching methods as can be found comparing the regression models among themselves (see Table 11 and Table 12).

In contrast to the regression models, the statistical matching models where values for the variable equivalised household income are imputed do even better regarding the bias for Austria than those where the factor at-risk-of-poverty is imputed.

Supplementary in Figure 4 the bias is plotted on the example of one small and one large state, namely Burgenland and Lower Austria, in order to gain a better view.

The best models concerning the absolute value of the bias for Austria and the nine states among all used methods can be found in Table 13.

Austria	logistic mixed regression model (e)(ii)
Burgenland	sequential random hot deck model (g)(ii)
Lower Austria	logistic regression model (c)(ii)
Vienna	logistic mixed regression model (e)(ii)
Carinthia	weighted random hot deck model (h)(iv)
Styria	random hot deck model (f)(i)
Upper Austria	weighted random hot deck model (h)(i)
Salzburg	sequential random hot deck model (g)(iv)
Tyrol	sequential random hot deck model (g)(iv)
Vorarlberg	logistic mixed regression model (e)(i)

Table 13: Models with the smallest absolute value of the bias for Austria and the nine states.

So in 4 of 10 cases the regression methods do better than the statistical matching methods. Maybe it is remarkable, that the sequential random hot deck models emerge as best model 3 times. But this result may be random and therefore has to be handled with caution.

4.3.3. Variance

On the one hand the variance is computed of the at-risk-of-poverty rates for Austria and the nine states that are estimated on the basis of the drawn EU-SILC samples, see also

47

Equation 5 and Equation 6. On the other hand for Austria and every of the nine regions it is calculated on basis of the estimated values in the drawn Micro Census data sets for every model seperately.

Concerning the variances calculated from the repeated sampling of EU-SILC samples it is conspicuously, that the estimations for the small states are more inaccurately than those for the bigger ones (see Table 14). The following order of the states corresponds to the variance in ascending order: Lower Austria, Upper Austria, Vienna, Styria, Tyrol, Salzburg, Carinthia, Vorarlberg, Burgenland. The standard deviation of the states ranges from 0.8361802% (Lower Austria) to 2.6507095% (Burgenland). For Austria the variance received from the EU-SILC samples is 0.4310486%.

	Austria	Burgenland	Lower Austria	Vienna	Carinthia
SD_S	0.431049	2.650709	0.836180	1.032131	1.785996

	Styria	Upper Austria	Salzburg	Tyrol	Vorarlberg
SD_S	1.093533	0.911877	1.503958	1.452394	2.184948

Table 14: Standard deviation on basis of the EU-SILC samples [in %].

Comparing the variance of the direct estimator with the variance of the model estimators it can be noticed that the variance for Austria is not undercut by any of the different models, see Table 14 and Table 15. This is due to the nature of these methods that should allow for smaller variances on regional level but the Micro Census data set is not that much bigger that this gets effective.

The comparison of the variances for the nine states shows - at least for the regression models - a different picture (see Table 14, Table 15 and Table 16):

Regression Models

Regarding as a start only the variances of the regression models for Burgenland, Salzburg, Tyrol or Vorarlberg it appears, that the logistic (mixed) regression models (c)(ii) and (e)(ii) have a larger variance than the direct estimator, but all the 8 other regression models have a smaller one. For Carinthia and Styria, respectively, the 4 models (a)(ii) and (d)(ii) as well as (c)(ii) and (e)(ii) have a larger variance, the other 6 regression models have a smaller one. Looking at Lower Austria or Upper Austria 5 regression models perform better in terms of the variance and 5 performs worse. For Vienna only 2 regression models do better than the direct estimator. So the advantage of the regression models can be found particularly in the estimations of the at-risk-of-poverty rates for the smaller states.

	Austria	Burgenland	Lower Aus.	Vienna	Carinthia
(a)(i) Linear Regression	0.643894	0.915786	0.805230	1.358796	0.830726
(a)(ii) Linear Regression	0.638561	1.735226	0.956579	1.348708	2.146662
(b)(i) Robust Lin.Reg.	0.455394	0.722326	0.648617	0.968362	0.689105
(b)(ii) Robust Lin.Reg.	0.451170	1.326309	0.741637	0.960381	1.108700
(c)(i) Logistic Reg.	0.519291	0.934264	0.729721	1.146252	0.862655
(c)(ii) Logistic Reg.	0.571466	2.911612	0.951076	1.204009	1.806076
(d)(i) Linear Mixed Reg.	0.641850	0.908475	0.799034	1.353469	0.828877
(d)(ii) Linear Mixed Reg.	0.638561	1.735226	0.956579	1.348708	2.146662
(e)(i) Logistic Mix.Reg.	0.608307	0.950801	0.871373	1.161652	0.941897
(e)(ii) Logistic Mix.Reg.	0.545751	2.842273	1.029850	1.158870	1.810080
(f)(i) Random Hot Deck	0.755507	3.651918	1.324693	1.723281	2.634175
(f)(ii) Random Hot Deck	0.708858	3.991881	1.349583	1.692774	2.751299
(f)(iii) Random Hot Deck	0.774372	3.850094	1.329489	1.580216	2.477023
(f)(iv) Random Hot Deck	0.709823	3.670203	1.309551	1.575223	2.848268
(g)(i) Sequential R.H.D.	0.851690	3.743581	1.835122	2.055633	2.805323
(g)(ii) Sequential R.H.D.	0.797239	3.750846	1.793504	2.044954	2.799569
(g)(iii) Sequential R.H.D.	0.690674	3.434548	1.271276	1.463391	2.452313
(g)(iv) Sequential R.H.D.	0.708064	3.480521	1.299959	1.491434	2.538129
(h)(i) Weighted R.H.D.	0.565343	2.799847	1.066875	1.242310	2.011388
(h)(ii) Weighted R.H.D.	0.527145	2.923010	0.992332	1.261184	2.014868
(h)(iii) Weighted R.H.D.	0.566486	2.717812	1.030578	1.197595	1.929699
(h)(iv) Weighted R.H.D.	0.522036	2.768013	1.031744	1.264949	1.907838

Table 15: Standard deviation for the several models calculated on basis of the Micro Census data sets [in %] - part 1: Austria, Burgenland, Lower Austria, Vienna, Carinthia.

Statistical Matching Methods

For the statistical matching methods the following holds again. No matter which state is considered, every of the statistical matching models results in a greater variance than the variance received from the EU-SILC samples. A remarkable fact is, that especially for the smaller states the weighted random hot deck models have a lower variance than the random hot deck or sequential random hot deck models.

Comparison: Regression Models and Statistical Matching Methods

As done for the bias and in order to gain a better view, the standard deviation is plotted on the example of one small and one large state, namely Burgenland and Lower Austria, in Figure 5. Due to the fact that both, the bias and the standard deviation, are desired to be small for one model and state, for the diagram the same states as above are selected deliberately.

	Styria	Upper Aus.	Salzburg	Tyrol	Vorarlberg
(a)(i) Linear Regression	0.829204	0.860510	0.835867	0.873983	1.103230
(a)(ii) Linear Regression	1.215587	1.020118	1.258288	1.269461	1.643167
(b)(i) Robust Lin.Reg.	0.627094	0.719934	0.675125	0.689733	0.822512
(b)(ii) Robust Lin.Reg.	0.888965	0.809586	0.911399	0.971983	1.323511
(c)(i) Logistic Reg.	0.791311	0.821617	0.893532	0.823094	0.934868
(c)(ii) Logistic Reg.	1.273035	1.088053	1.559762	1.529829	2.311870
(d)(i) Linear Mixed Reg.	0.826503	0.862319	0.836952	0.883440	1.108039
(d)(ii) Linear Mixed Reg.	1.215587	1.020118	1.258288	1.269461	1.643167
(e)(i) Logistic Mix.Reg.	0.926384	0.947702	0.959364	0.916694	1.054924
(e)(ii) Logistic Mix.Reg.	1.166038	1.060178	1.611479	1.508483	2.254460
(f)(i) Random Hot Deck	1.649661	1.469261	2.165907	2.096122	3.336729
(f)(ii) Random Hot Deck	1.696115	1.368381	2.215134	2.207942	3.434193
(f)(iii) Random Hot Deck	1.652477	1.408134	2.243846	2.255994	3.333223
(f)(iv) Random Hot Deck	1.890050	1.441082	2.215710	2.245211	3.294990
(g)(i) Sequential R.H.D.	2.315136	1.825121	2.449775	2.526301	3.253550
(g)(ii) Sequential R.H.D.	2.204776	1.766810	2.403209	2.513961	3.243946
(g)(iii) Sequential R.H.D.	1.644768	1.339885	1.975818	2.117027	2.889652
(g)(iv) Sequential R.H.D.	1.643382	1.391087	1.968525	2.157239	3.061581
(h)(i) Weighted R.H.D.	1.264378	1.096634	1.642210	1.663267	2.299018
(h)(ii) Weighted R.H.D.	1.321136	1.142284	1.609067	1.671752	2.423293
(h)(iii) Weighted R.H.D.	1.280591	0.998503	1.640461	1.695679	2.271055
(h)(iv) Weighted R.H.D.	1.314752	1.067349	1.618298	1.595950	2.375946

Table 16: Standard deviation for the several models calculated on basis of the Micro Census data sets [in %] - part 2: Styria, Upper Austria, Salzburg, Tyrol, Vorarlberg.

A possibility to evaluate the different methods and models is again the examination of the average of the nine states (see also Appendix A.2, Table 33). The average of the variances is lowest for the robust linear regression model (b)(i), followed by the logistic regression model (c)(i), the linear mixed regression model (d)(i), the linear regression model (a)(i) and the logistic mixed regression model (e)(i). These 5 models as well as the robust linear regression model (b)(ii), the linear regression model (a)(ii) and the linear mixed regression model (d)(ii) have a lower average than the direct estimator. The logistic (mixed) regression models (e)(ii) and (c)(ii) have already a higher average, but still a lower one than almost all the statistical matching models. Here the same picture as described above can be found: every of the four weighted random hot deck models have a lower average of variances than the random hot deck or sequential random hot deck models.

Which models emerge as the best models concerning the variance for Austria and the nine states among all used methods can be found in Table 17.

So in terms of the variance the robust linear regression models perform best quite clearly. Equally which state or Austria is considered, they do best in every case.

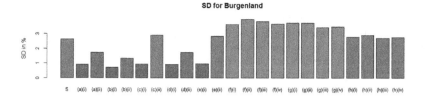

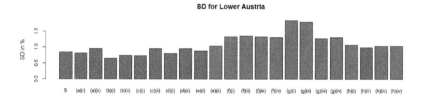

Figure 5: Bar chart of the standard deviation [in %] for the direct estimator (calculated on basis of the EU-SILC data sets)(denoted by S) and the several models (calculated on basis of the Micro Census data sets) using the example of Burgenland and Lower Austria.

Austria	robust linear regression model (b)(ii)
Burgenland	robust linear regression model (b)(i)
Lower Austria	robust linear regression model (b)(i)
Vienna	robust linear regression model (b)(ii)
Carinthia	robust linear regression model (b)(i)
Styria	robust linear regression model (b)(i)
Upper Austria	robust linear regression model (b)(i)
Salzburg	robust linear regression model (b)(i)
Tyrol	robust linear regression model (b)(i)
Vorarlberg	robust linear regression model (b)(i)

Table 17: Models with the smallest variance for Austria and the nine states.

4.3.4. Mean Squared Error

For the calculation of the MSE for both, the EU-SILC and (enhanced) Micro Census data sets, the respective bias and variances get combined for Austria and every of the nine states, respectively (see also Equation 7).

51

Looking as a start at the MSE calculated for the estimations based on the EU-SILC data sets it is remarkable that there results the same ascending order as for the variances of the EU-SILC estimations, that is to say Lower Austria, Upper Austria, Vienna, Styria, Tyrol, Salzburg, Carinthia, Vorarlberg, Burgenland (see Table 18). Again the direct estimator performs better for the bigger states. The MSE for the states ranges from 7.052380e-05 (Lower Austria) to 7.040865e-04 (Burgenland). The estimation for Austria has the lowest MSE with 2.018856e-05.

	Austria	Burgenland	Lower Austria	Vienna	Carinthia
MSES	2.019e-05	7.041e-04	7.052e-05	1.145e-04	3.191e-04

	Styria	Upper Austria	Salzburg	Tyrol	Vorarlberg
MSES	1.206e-04	8.329e-05	2.269e-04	2.118e-04	4.868e-04

Table 18: MSE on basis of the EU-SILC samples.

	Austria	Burgenland	Lower Aus.	Vienna	Carinthia
(a)(i) Linear Regression	6.726e-04	0.008834	3.131e-04	2.668e-04	0.008834
(a)(ii) Linear Regression	6.167e-04	0.011080	3.021e-04	2.453e-04	0.003153
(b)(i) Robust Lin.Reg.	0.002076	0.012406	0.001071	0.001649	0.010793
(b)(ii) Robust Lin.Reg.	0.002007	0.014500	0.001010	0.001575	0.009165
(c)(i) Logistic Reg.	2.829e-05	0.003276	2.755e-04	1.363e-04	0.002927
(c)(ii) Logistic Reg.	3.410e-05	8.487e-04	9.051e-05	1.507e-04	3.266e-04
(d)(i) Linear Mixed Reg.	6.325e-04	0.008801	3.073e-04	2.227e-04	0.007405
(d)(ii) Linear Mixed Reg.	6.167e-04	0.011080	3.021e-04	2.453e-04	0.003153
(e)(i) Logistic Mix.Reg.	1.016e-04	0.002378	6.255e-04	1.396e-04	0.002115
(e)(ii) Logistic Mix.Reg.	3.107e-05	8.099e-04	1.063e-04	1.381e-04	3.283e-04
(f)(i) Random Hot Deck	1.113e-04	0.001337	1.761e-04	3.051e-04	6.944e-04
(f)(ii) Random Hot Deck	5.330e-05	0.001594	1.842e-04	2.992e-04	7.587e-04
(f)(iii) Random Hot Deck	1.137e-04	0.001484	1.787e-04	2.661e-04	6.172e-04
(f)(iv) Random Hot Deck	5.338e-05	0.001353	1.722e-04	2.608e-04	8.117e-04
(g)(i) Sequential R.H.D.	5.067e-04	0.001470	7.143e-04	6.759e-04	0.001287
(g)(ii) Sequential R.H.D.	1.185e-04	0.001407	5.337e-04	4.498e-04	9.455e-04
(g)(iii) Sequential R.H.D.	9.658e-05	0.001180	1.652e-04	2.369e-04	6.092e-04
(g)(iv) Sequential R.H.D.	5.450e-05	0.001218	1.749e-04	2.562e-04	6.458e-04
(h)(i) Weighted R.H.D.	8.894e-05	7.862e-04	1.144e-04	1.633e-04	4.050e-04
(h)(ii) Weighted R.H.D.	2.996e-05	8.637e-04	9.897e-05	1.672e-04	4.076e-04
(h)(iii) Weighted R.H.D.	9.485e-05	7.415e-04	1.074e-04	1.506e-04	3.734e-04
(h)(iv) Weighted R.H.D.	2.869e-05	7.682e-04	1.066e-04	1.683e-04	3.643e-04

Table 19: MSE for the several models calculated on basis of the Micro Census data sets - part 1: Austria, Burgenland, Lower Austria, Vienna, Carinthia.

The MSE for Austria is not untercut by any of the models, see Table 18 and Table 19. Nor for the states Burgenland, Lower Austria, Vienna, Carinthia or Tyrol any method succeed in resulting in a lower MSE than the direct estimator. However, for Styria and Salzburg the estimations of the logistic regression model (c)(i) have a lower MSE. Further the estimator for Upper Austria based on the linear regression model (a)(i) or the linear mixed regression model (d)(i) untercuts the direct estimator for Upper Austria. Moreover the logistic regression model (c)(i) as well as the logistic mixed regression model (e)(i) performs better than the direct estimator regarding Vorarlberg. For details see Table 18, Table 19 and Table 20.

None of the statistical matching models has a lower MSE than the direct estimator, neither for Austria nor for any of the states. As already seen for the variances it appears that especially for the smaller states the weighted random hot deck models have a lower MSE than the random hot deck or sequential random hot deck models, see Table 19 and Table 20.

	Styria	Upper Aus.	Salzburg	Tyrol	Vorarlberg
(a)(i) Linear Regression	9.222e-04	7.960e-05	7.538e-04	0.002978	0.001302
(a)(ii) Linear Regression	2.527e-04	1.403e-04	0.001750	0.003405	0.004527
(b)(i) Robust Lin.Reg.	0.002346	2.766e-04	0.001799	0.004841	0.003734
(b)(ii) Robust Lin.Reg.	8.846e-04	5.897e-04	0.002986	0.004892	0.005717
(c)(i) Logistic Reg.	6.406e-05	8.227e-04	1.640e-04	4.875e-04	1.832e-04
(c)(ii) Logistic Reg.	1.642e-04	1.187e-04	2.434e-04	2.370e-04	5.468e-04
(d)(i) Linear Mixed Reg.	9.330e-04	8.242e-05	7.154e-04	0.002909	0.001201
(d)(ii) Linear Mixed Reg.	2.527e-04	1.403e-04	0.001750	0.003405	0.004527
(e)(i) Logistic Mix.Reg.	1.744e-04	0.001425	4.406e-04	2.381e-04	1.114e-04
(e)(ii) Logistic Mix.Reg.	1.377e-04	1.131e-04	2.610e-04	2.276e-04	5.146e-04
(f)(i) Random Hot Deck	2.721e-04	2.164e-04	4.692e-04	4.396e-04	0.001122
(f)(ii) Random Hot Deck	2.877e-04	1.899e-04	4.907e-04	4.930e-04	0.001206
(f)(iii) Random Hot Deck	2.734e-04	1.987e-04	5.040e-04	5.097e-04	0.001112
(f)(iv) Random Hot Deck	3.630e-04	2.084e-04	4.928e-04	5.049e-04	0.001100
(g)(i) Sequential R.H.D.	8.070e-04	4.982e-04	7.406e-04	7.144e-04	0.001216
(g)(ii) Sequential R.H.D.	5.621e-04	3.452e-04	6.027e-04	6.321e-04	0.001060
(g)(iii) Sequential R.H.D.	2.708e-04	1.803e-04	3.908e-04	4.492e-04	8.427e-04
(g)(iv) Sequential R.H.D.	2.723e-04	1.935e-04	3.875e-04	4.654e-04	9.567e-04
(h)(i) Weighted R.H.D.	1.608e-04	1.203e-04	2.701e-04	2.779e-04	5.345e-04
(h)(ii) Weighted R.H.D.	1.754e-04	1.307e-04	2.603e-04	2.809e-04	6.042e-04
(h)(iii) Weighted R.H.D.	1.657e-04	9.973e-05	2.702e-04	2.913e-04	5.294e-04
(h)(iv) Weighted R.H.D.	1.741e-04	1.140e-04	2.626e-04	2.551e-04	5.773e-04

Table 20: MSE for the several models calculated on basis of the Micro Census data sets - part 2: Styria, Upper Austria, Salzburg, Tyrol, Vorarlberg.

In conclusion the mean squared error is plotted in Figure 6 using the same states as above, Burgenland and Lower Austria.

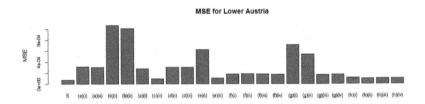

Figure 6: Bar chart of the mean squared error for the direct estimator (calculated on basis of the EU-SILC data sets)(denoted by S) and the several models (calculated on basis of the Micro Census data sets) using the example of Burgenland and Lower Austria.

Considering the average of the nine states for the direct estimator and the models (see Appendix A.2, Table 34) someone realizes again that the direct estimator performs best in terms of the mean squared error. It is followed by the logistic mixed regression model (e)(ii), the logistic regression model (c)(ii), the weighted random hot deck models (h)(iii), (h)(iv), (h)(i) and (h)(ii). In the further order the sequential random hot deck models and the random hot deck models appear and then the other regression models follow.

Austria	logistic regression model (c)(i)
Burgenland	weighted random hot deck model (h)(iii)
Lower Austria	logistic regression model (c)(ii)
Vienna	logistic regression model (c)(i)
Carinthia	logistic regression model (c)(ii)
Styria	logistic regression model (c)(i)
Upper Austria	linear regression model (a)(i)
Salzburg	logistic regression model (c)(i)
Tyrol	logistic mixed regression model (e)(ii)
Vorarlberg	logistic mixed regression model (e)(i)

Table 21: Models with the smallest MSE for Austria and the nine states

Interesting is the result of the search for the best model concerning the MSE for Austria and the nine states among all used methods. The finding is summarized in Table 21. Only in one case the statistical matching methods can keep up with the regression models, namely for Burgenland. In other respects the logistic (mixed) regression models emerge as best model many times (more precisely in 8 of 10 cases).

4.3.5. Bias Corrected Version

Using the mean of the bootstrap replicates (of all drawn EU-SILC and Micro Census samples, respectively) for the computation of the bias instead of the mean of the at-risk-of-poverty rates obtained only from the repeated sampling of EU-SILC and Micro Census data sets, respectively, an in some sense "corrected" version of the bias is calculated, see also Equation 12 and Equation 13.

The new bias obtained from the EU-SILC samples is a little bit higher than the one described above (see Table 22).

	Austria	Burgenland	Lower Austria	Vienna	Carinthia
BiasBS (Eq. 13)	0.17868	0.22340	0.11464	0.33785	0.12097

	Styria	Upper Austria	Salzburg	Tyrol	Vorarlberg
BiasBS (Eq. 13)	0.14546	0.07793	0.12118	0.15468	0.39951

Table 22: Corrected bias on basis of the EU-SILC samples [in %].

Regression Models

The results for the regression models are very similar (see Table 23 and Table 24):

The (robust) linear regression models perform worse than the direct estimator and the logistic regression models do better than (robust) linear regression. Furthermore the same pattern regarding the mixed regression models can be observed.

Also the circumstance that the linear (mixed) regression models result mostly in negative figures and the logistic (mixed) regression models in positive ones is preserved.

As best model concerning the corrected bias for Austria among the regression models emerges again the logistic mixed regression model (e)(ii) and again it results in a smaller bias than the direct estimator. The previously calculated absolute value of the bias of the logistic mixed regression model (e)(ii) has been the smallest of all regression models for the states Vienna and Tyrol - this fact changes in this setup: the logistic mixed regression model (e)(ii) performs best for the states Lower Austria, Carinthia, Styria, Upper Austria and Tyrol. The logistic regression model (c)(ii) does best for Burgenland and Salzburg and the logistic mixed regression model (e)(i) performs again best for Vorarlberg, but also for Vienna. The absolute value of the smallest bias among the regression models is

	Austria	Burgenland	Lower Aus.	Vienna	Carinthia
(a)(i) Linear Regression	-2.24832	-9.03837	-1.28876	-0.67062	-8.23425
(a)(ii) Linear Regression	-2.11000	-9.93333	-1.16961	-0.57087	-4.75556
(b)(i) Robust Lin.Reg.	-5.10056	-11.55437	-3.61510	-4.82066	-10.78143
(b)(ii) Robust Lin.Reg.	-5.02027	-12.18061	-3.52358	-4.75545	-9.94436
(c)(i) Logistic Reg.	0.17178	-5.51241	1.56089	0.27791	-5.27414
(c)(ii) Logistic Reg.	0.16916	0.28062	0.08987	0.27642	0.13974
(d)(i) Linear Mixed Reg.	-2.17778	-9.02706	-1.27545	-0.44194	-8.25622
(d)(ii) Linear Mixed Reg.	-2.11000	-9.93333	-1.16961	-0.57087	-4.75556
(e)(i) Logistic Mix.Reg.	0.82422	-4.83336	2.35466	0.25851	-4.54477
(e)(ii) Logistic Mix.Reg.	0.16721	0.28641	0.08288	0.27892	0.13058
(f)(i) Random Hot Deck	0.81255	0.22066	0.11060	0.33572	0.13572
(f)(ii) Random Hot Deck	0.23722	0.33165	0.14553	0.40815	0.19131
(f)(iii) Random Hot Deck	0.82660	0.24171	0.12378	0.34667	0.13745
(f)(iv) Random Hot Deck	0.23477	0.29294	0.15816	0.40916	0.19494
(g)(i) Sequential R.H.D.	1.80835	0.66296	1.49741	1.41561	1.61825
(g)(ii) Sequential R.H.D.	0.54154	-0.07693	1.05475	0.53751	0.79139
(g)(iii) Sequential R.H.D.	0.71980	0.11153	0.17803	0.45593	-0.20306
(g)(iv) Sequential R.H.D.	0.24059	0.27100	0.25186	0.60374	-0.02985
(h)(i) Weighted R.H.D.	0.82057	0.23697	0.12080	0.33336	0.12970
(h)(ii) Weighted R.H.D.	0.20577	0.25912	0.13233	0.37166	0.15800
(h)(iii) Weighted R.H.D.	0.81884	0.21729	0.12554	0.33727	0.10868
(h)(iv) Weighted R.H.D.	0.20331	0.24254	0.13650	0.37672	0.13925

Table 23: Corrected bias for the several models calculated on basis of the Micro Census data sets [in %] - part 1: Austria, Burgenland, Lower Austria, Vienna, Carinthia.

lower than the respective bias obtained from the EU-SILC samples for Lower Austria, Vienna, Styria, Salzburg and Vorarlberg. For details have a look at Table 22, Table 23 and Table 24.

The ordering of the models regarding the average of the bias of the nine states (see Appendix A.2, Table 32) changes also a little bit: Now the logistic mixed regression model (e)(i) performs best among the regression models, followed by the logistic mixed regression model (e)(ii), the logistic regression model (c)(ii) and then the logistic regression model (c)(i).

Taking the absolute values for the computation of the average (see Appendix A.2, Table 32) there can be obtained the same ordering as above: the logistic mixed regression model (e)(ii) performs best, followed by the logistic regression model (c)(ii), then by the logistic regression model (c)(i) and the logistic mixed regression model (e)(i).

	Styria	Upper Aus.	Salzburg	Tyrol	Vorarlberg
(a)(i) Linear Regression	-2.67989	0.45942	-2.31085	-5.07554	-3.15039
(a)(ii) Linear Regression	-0.78260	-0.36273	-3.65577	-5.34478	-6.06326
(b)(i) Robust Lin.Reg.	-5.25685	-2.10845	-4.65652	-7.37730	-6.75663
(b)(ii) Robust Lin.Reg.	-3.30143	-2.93363	-5.85351	-7.23120	-8.08329
(c)(i) Logistic Reg.	0.16782	2.77294	1.04055	-2.00128	-0.97900
(c)(ii) Logistic Reg.	0.13741	0.10973	0.10888	0.17255	0.38940
(d)(i) Linear Mixed Reg.	-2.69027	0.51239	-2.24374	-5.01850	-3.01849
(d)(ii) Linear Mixed Reg.	-0.78260	-0.36273	-3.65577	-5.34478	-6.06326
(e)(i) Logistic Mix.Reg.	0.97015	3.70033	1.90834	-1.24644	-0.01836
(e)(ii) Logistic Mix.Reg.	0.13122	0.10971	0.11744	0.16173	0.40179
(f)(i) Random Hot Deck	0.15045	0.06994	0.11126	0.13793	0.37676
(f)(ii) Random Hot Deck	0.21258	0.13186	0.17234	0.20140	0.49105
(f)(iii) Random Hot Deck	0.14372	0.08545	0.15023	0.14912	0.40150
(f)(iv) Random Hot Deck	0.21826	0.12881	0.15721	0.17176	0.47145
(g)(i) Sequential R.H.D.	1.38562	0.86757	0.95313	0.81558	1.18801
(g)(ii) Sequential R.H.D.	0.69308	0.24077	0.35970	0.03881	0.31848
(g)(iii) Sequential R.H.D.	0.03461	-0.03498	-0.08318	-0.07824	0.36171
(g)(iv) Sequential R.H.D.	0.15602	0.06967	0.00811	0.03702	0.55101
(h)(i) Weighted R.H.D.	0.15033	0.07442	0.13027	0.14961	0.40239
(h)(ii) Weighted R.H.D.	0.17305	0.09644	0.14206	0.18606	0.44722
(h)(iii) Weighted R.H.D.	0.14410	0.07520	0.13267	0.14385	0.42007
(h)(iv) Weighted R.H.D.	0.16990	0.09467	0.14590	0.16515	0.44199

Table 24: Corrected bias for the several models calculated on basis of the Micro Census data sets [in %] - part 2: Styria, Upper Austria, Salzburg, Tyrol, Vorarlberg.

Statistical Matching Methods

Comparing the corrected bias for Austria among the statistical matching models one will notice that again the weighted random hot deck model (h)(iv) has the smallest absolute value, see Table 23. However, now the bias of the weighted random hot deck model (h)(iv) is larger than the bias of the direct estimator for Austria (see Table 22 and Table 23).

Comparing the statistical matching models for the several states, similar results are obtained (see Table 23 and Table 24): Again for Burgenland the sequential random hot deck model (g)(ii) performs best and the absolute value of the bias for Salzburg and Tyrol is again smallest for the estimations based on the sequential random hot deck model (g)(iv). In contrast to above the best statistical matching estimator in terms of the bias for Lower Austria is the random hot deck model (f)(i) (instead of (h)(iv)), in terms of the bias for Vienna is the weighted random hot deck model (h)(i) (instead of (h)(iii)), in terms of the bias for Carinthia it is the sequential random hot deck model (g)(iv) (instead of (h)(iv)), in terms of the bias for Styria and Upper Austria, respectively, it is the sequential random hot deck model (g)(iii) (instead of (f)(i) and (h)(i), respectively) and in terms of the bias for Vorarlberg it is the sequential random hot deck model (g)(ii) (instead of (f)(iii)).

Every of these 9 (best) methods results in a absolute value of the bias comparable in quality to the direct estimator (see also Table 22).

Looking again at the average of the nine states (see Appendix A.2, Table 32) the ordering changes a little bit: sequential random hot deck model (g)(iii) performs best, followed by the random hot deck model (f)(i) and then by the weighted random hot deck models (h)(iii) and (h)(i). Taking the absolute values the result looks again different: the sequential random hot deck model (g)(iii) does best, followed by the random hot deck model (f)(i), the weighted random hot deck models (h)(iii) and (h)(i), then by the random hot deck model (f)(iii), the weighted random hot deck models (h)(iv), (h)(ii) and the sequential random hot deck model (g)(iv).

Comparison: Regression Models and Statistical Matching Methods

The models that emerge as the best models concerning the absolute value of the corrected bias for Austria and the nine states among all used methods can be found in Table 25. There is a column indicating which model does best regarding the bias calculated and evaluated in Chapter 4.3.2.

	corrected bias	"old" bias
Austria	logistic mixed regression model (e)(ii)	(e)(ii)
Burgenland	sequential random hot deck model (g)(ii)	(g)(ii)
Lower Austria	logistic mixed regression model (e)(ii)	(c)(ii)
Vienna	logistic mixed regression model (e)(i)	(e)(ii)
Carinthia	sequential random hot deck model (g)(iv)	(h)(iv)
Styria	sequential random hot deck model (g)(iii)	(f)(i)
Upper Austria	sequential random hot deck model (g)(iii)	(h)(i)
Salzburg	sequential random hot deck model (g)(iv)	(g)(iv)
Tyrol	sequential random hot deck model (g)(iv)	(g)(iv)
Vorarlberg	logistic mixed regression model (e)(i)	(e)(i)

Table 25: Models with the smallest absolute value of the (corrected) bias for Austria and the nine states.

In 4 of 10 cases the regression methods do better than the statistical matching methods. It is remarkable, that the sequential random hot deck models emerge as best model 6 times.

Comparing the above introduced bias and the corrected version in 5 of 10 cases there results the same models.

Regarding the absolute values of the two different versions of the bias for every model for Austria/every state it is prominent, that the linear (mixed) regression models (a)(i), (a)(ii), (d)(i) and (d)(ii) have a lower corrected bias almost always, only for Lower Austria the models (a)(i) and (d)(i) result in a higher new bias. For the other regression models in only 5 occurrences the absolute value of the corrected bias is lower than the old one (All in all, 55 times the absolute bias introduced in Chapter 4.3.2 is lower). Similar results are obtained looking at the statistical matching methods: The sequential random hot deck models (especially model (g)(i), (g)(ii) and (g)(iii)) have more often than not a lower new bias, but the random hot deck models and the weighted random hot deck models result only in 10 of 80 cases in a lower corrected bias.

This corrected bias can now be used for the computation of the MSE for once the EU-SILC samples and once the Micro Census data sets in order to get a bias corrected version of the MSE.

	Austria	Burgenland	Lower Austria	Vienna	Carinthia
MSEBS	2.177e-05	7.076e-04	7.123e-05	1.179e-04	3.204e-04

	Styria	Upper Austria	Salzburg	Tyrol	Vorarlberg
MSEBS	1.217e-04	8.376e-05	2.277e-04	2.133e-04	4.934e-04

Table 26: Bias corrected MSE on basis of the EU-SILC samples.

Looking at the bias corrected version of the MSE (see Table 26) calculated for the estimations based on the EU-SILC data sets someone notices that there results the same ascending order as obtained before for the variances or the MSE calculated in Chapter 4.3.4, i.e. Lower Austria followed by Upper Austria, Vienna, Styria, Tyrol, Salzburg, Carinthia, Vorarlberg and Burgenland. The bias corrected MSE for the states ranges from 7.123386e-05 (Lower Austria) to 7.076170e-04 (Burgenland). The estimation for Austria has the lowest MSE with 2.177282e-05. The bias corrected mean squared errors are a little bit higher than the values of the MSE dealt with in Chapter 4.3.2 based on the EU-SILC data sets.

Also the bias corrected MSE for Austria is not undercut by any of the models, see Table 26 and Table 27.

Concerning the bias corrected MSE for the states nearly the same dimensions as for the MSE of Chapter 4.3.4 result (see Table 26, see Table 27 and Table 28):

Again nor for the states Burgenland, Lower Austria, Vienna, Carinthia or Tyrol any method succeed in resulting in a lower MSE than the direct estimator. Afresh for Styria and Salzburg the estimations of the logistic regression model (c)(i) have a lower MSE and the logistic regression model (c)(i) as well as the logistic mixed regression model (e)(i) performs better than the direct estimator regarding Vorarlberg. The only change is, that for Upper Austria also none of the models can undercut the bias corrected MSE of the direct estimator.

Again none of the statistical matching models has a lower MSE than the direct estimator, neither for Austria nor for any of the states. As already seen and described further

	Austria	Burgenland	Lower Aus.	Vienna	Carinthia
(a)(i) Linear Regression	5.470e-04	0.008253	2.309e-04	2.296e-04	0.006849
(a)(ii) Linear Regression	4.860e-04	0.006849	2.283e-04	2.145e-04	0.006849
(b)(i) Robust Lin.Reg.	0.002622	0.013403	0.001349	0.002418	0.011671
(b)(ii) Robust Lin.Reg.	0.002541	0.015013	0.001297	0.002354	0.010012
(c)(i) Logistic Reg.	2.992e-05	0.003126	2.969e-04	1.391e-04	0.002856
(c)(ii) Logistic Reg.	3.552e-05	8.556e-04	9.126e-05	1.526e-04	3.281e-04
(d)(i) Linear Mixed Reg.	5.155e-04	0.008231	2.265e-04	2.027e-04	0.006885
(d)(ii) Linear Mixed Reg.	4.860e-04	0.010168	2.283e-04	2.145e-04	0.002722
(e)(i) Logistic Mix.Reg.	1.049e-04	0.002427	6.304e-04	1.416e-04	0.002154
(e)(ii) Logistic Mix.Reg.	3.258e-05	8.161e-04	1.067e-04	1.421e-04	3.293e-04
(f)(i) Random Hot Deck	1.231e-04	0.001339	1.767e-04	3.082e-04	6.957e-04
(f)(ii) Random Hot Deck	5.588e-05	0.001605	1.843e-04	3.032e-04	7.606e-04
(f)(iii) Random Hot Deck	1.283e-04	0.001488	1.783e-04	2.617e-04	6.155e-04
(f)(iv) Random Hot Deck	5.590e-05	0.001356	1.740e-04	2.649e-04	8.151e-04
(g)(i) Sequential R.H.D.	3.996e-04	0.001445	5.610e-04	6.230e-04	0.001049
(g)(ii) Sequential R.H.D.	9.289e-05	0.001407	4.329e-04	4.471e-04	8.464e-04
(g)(iii) Sequential R.H.D.	9.951e-05	0.001181	1.648e-04	2.349e-04	6.055e-04
(g)(iv) Sequential R.H.D.	5.592e-05	0.001219	1.753e-04	2.589e-04	6.443e-04
(h)(i) Weighted R.H.D.	9.930e-05	7.895e-04	1.153e-04	1.654e-04	4.063e-04
(h)(ii) Weighted R.H.D.	3.202e-05	8.611e-04	1.002e-04	1.729e-04	4.085e-04
(h)(iii) Weighted R.H.D.	9.914e-05	7.434e-04	1.078e-04	1.548e-04	3.736e-04
(h)(iv) Weighted R.H.D.	3.139e-05	7.721e-04	1.083e-04	1.742e-04	3.659e-04

Table 27: Bias corrected MSE for the several models calculated on basis of the Micro Census data sets - part 1: Austria, Burgenland, Lower Austria, Vienna, Carinthia.

up for the variances and the old MSE it appears, that especially for the smaller states the weighted random hot deck models have a lower MSE than the random hot deck or sequential random hot deck models.

The average of the nine states for the direct estimator and every model (see Appendix A.2, Table 34) yields the same ordering as for the old MSE with the exception of the change of the models (c)(ii) and (h)(iii) as well as the change of the models (g)(i) and (e)(i). So the direct estimator performs best in terms of the bias corrected mean squared error and it is followed by the logistic mixed regression model (e)(ii), the weighted random hot deck model (h)(iii), the logistic regression model (c)(ii), the weighted random hot deck models (h)(iv), (h)(i) and (h)(ii). In the further order the sequential random hot deck models and the random hot deck models are found again and then the other regression models follow.

The result of the search for the best model concerning the bias corrected MSE for Austria and the nine states among all used method is summarized in Table 29. Moreover a column indicates which model emerges as the best one concerning the old MSE calculated

	Styria	Upper Aus.	Salzburg	Tyrol	Vorarlberg
(a)(i) Linear Regression	7.869e-04	9.515e-05	6.039e-04	0.002652	0.001114
(a)(ii) Linear Regression	2.090e-04	1.172e-04	0.001495	0.003018	0.003946
(b)(i) Robust Lin.Reg.	0.002803	4.964e-04	0.002214	0.005490	0.004633
(b)(ii) Robust Lin.Reg.	0.001169	9.262e-04	0.003509	0.005323	0.006709
(c)(i) Logistic Reg.	6.543e-05	8.364e-04	1.881e-04	4.683e-04	1.832e-04
(c)(ii) Logistic Reg.	1.639e-04	1.196e-04	2.445e-04	2.370e-04	5.496e-04
(d)(i) Linear Mixed Reg.	7.921e-04	1.006e-04	5.735e-04	0.002597	0.001034
(d)(ii) Linear Mixed Reg.	2.090e-04	1.172e-04	0.001495	0.003018	0.003946
(e)(i) Logistic Mix.Reg.	1.799e-04	0.001459	4.562e-04	2.394e-04	1.113e-04
(e)(ii) Logistic Mix.Reg.	1.377e-04	1.136e-04	2.611e-04	2.302e-04	5.244e-04
(f)(i) Random Hot Deck	2.744e-04	2.164e-04	4.704e-04	4.413e-04	0.001128
(f)(ii) Random Hot Deck	2.922e-04	1.890e-04	4.937e-04	4.916e-04	0.001203
(f)(iii) Random Hot Deck	2.751e-04	1.990e-04	5.057e-04	5.112e-04	0.001127
(f)(iv) Random Hot Deck	3.620e-04	2.093e-04	4.934e-04	5.070e-04	0.001108
(g)(i) Sequential R.H.D.	7.280e-04	4.084e-04	6.910e-04	7.047e-04	0.001200
(g)(ii) Sequential R.H.D.	5.341e-04	3.180e-04	5.905e-04	6.322e-04	0.001062
(g)(iii) Sequential R.H.D.	2.706e-04	1.797e-04	3.911e-04	4.488e-04	8.481e-04
(g)(iv) Sequential R.H.D.	2.725e-04	1.940e-04	3.875e-04	4.655e-04	9.677e-04
(h)(i) Weighted R.H.D.	1.621e-04	1.208e-04	2.714e-04	2.789e-04	5.447e-04
(h)(ii) Weighted R.H.D.	1.775e-04	1.314e-04	2.609e-04	2.829e-04	6.072e-04
(h)(iii) Weighted R.H.D.	1.661e-04	1.003e-04	2.709e-04	2.896e-04	5.334e-04
(h)(iv) Weighted R.H.D.	1.757e-04	1.148e-04	2.640e-04	2.574e-04	5.840e-04

Table 28: Bias corrected MSE for the several models calculated on basis of the Micro Census data sets - part 2: Styria, Upper Austria, Salzburg, Tyrol, Vorarlberg.

in Chapter 4.3.4.

	bias corrected MSE	"old" MSE
Austria	logistic regression model (c)(i)	(c)(i)
Burgenland	weighted random hot deck model (h)(iii)	(h)(iii)
Lower Austria	logistic regression model (c)(ii)	(c)(ii)
Vienna	logistic regression model (c)(i)	(c)(i)
Carinthia	logistic regression model (c)(ii)	(c)(ii)
Styria	logistic regression model (c)(i)	(c)(i)
Upper Austria	linear regression model (a)(i)	(a)(i)
Salzburg	logistic regression model (c)(i)	(c)(i)
Tyrol	logistic mixed regression model (e)(ii)	(e)(ii)
Vorarlberg	logistic mixed regression model (e)(i)	(e)(i)

Table 29: Models with the smallest (bias corrected) MSE for Austria and the nine states.

The same models re-emerge as best models for Austria and the 9 states. In other words again only in one case the statistical matching methods can keep up with the regression models, namely for Burgenland and in other respects the logistic (mixed) regression models emerge as best model many times (more precisely in 8 of 10 cases).

5. Conclusion

In conclusion it doesn't exist only one method performing by far best in every sense, and using the Micro Census gives hardly any improvement over the direct estimator on EU-SILC in terms of the mean squared error.

However, when comparing the methods used for the estimation of the at-risk-at-poverty rate based on the Micro Census data, some interesting conclusions can be made. In addition we have shown that the variance decreases for various regression models for small states in comparison to the direct estimator, but the bias may increase in such a manner, so that the mean squared error is still lowest for the direct estimator.

In more detail, concerning the bias, the logistic (mixed) regression models stand out in a positive manner among the regression models. The statistical matching methods differ not much in their performance. Apart from that it is worth noting that for the states Burgenland, Salzburg and Tyrol one of the sequential random hot deck models does best (once model (g)(ii) and twice model (g)(iv)). For Austria the logistic mixed regression model (e)(ii) performs best. For 7 of the 9 states the statistical matching methods do better than the regression methods.

Regarding the variances in general the regression models have better performance than the statistical matching methods. But the two models (c)(ii) and (e)(ii) proved to be quite good in terms of the bias have a higher variance. The robust linear regression models perform best quite clearly. This is in contrast to the result yield from the evaluation of the bias. Among the statistical matching methods, especially for the smaller states, the weighted random hot deck models have a lower variance than the random hot deck or sequential random hot deck models.

The combination of the bias and the variance in the MSE shows the problem described at the beginning of the thesis: due to the sample sizes a clear ascending order results. The smaller the number of respondents in the states, the larger the uncertainty. The ordering is Lower Austria, Upper Austria, Vienna, Styria, Tyrol, Salzburg, Carinthia, Vorarlberg, Burgenland.

Looking at the model with the lowest MSE for every state seperately only in one case the statistical matching methods can keep up with the regression models, namely for Burgenland. In other respects the logistic regression models and the logistic mixed regression models emerge as best model many times (more precisely in 8 of 10 cases). But for Austria and for 5 of the 9 states the MSE of the direct estimator can't be undercut by any of the models. For the other 4 states only one or two models perform better than the direct estimator. This may result because the available variables are not the most appropriate variables for the estimation of the equivalised household income or the variable at-risk-of-poverty. Furthermore the majority of the variables are factors (with only a few categories) and not continuous variables. Another reason is that the Micro Census incorporates "only" about 23000 households, but Austria has about 3.65 million households [see, e.g., Moser et al., 2013]. Thus using census data or register data the bias would possibly remain the same, but the variance and hence the mean squared error would decrease.

Considering the average of the nine states for the direct estimator and every model one can realize that the direct estimator performs best in terms of the mean squared error. It is followed by the logistic mixed regression model (e)(ii), the logistic regression model (c)(ii) and the 4 weighted random hot deck models. Further the sequential random hot deck models and the random hot deck models follow in quality before the other regression models.

The reliability of the bootstrapping can be verified for every of the methods and models. Furthermore the bias corrected versions yield similar results.

An application of additional methods and models may be future work. On the one side it is interesting whether the results will change by increasing the number of bootstrap replicates or the number of sample drawing and on the other side other models and methods may perform better.

Furthermore, another interesting approach would be to apply the described and used methods and models for other countries and data. Due to the fact that it is worked with design-based simulation, the adaption would take place straightforwardly. The analysis of the change caused by the usage of several sample designs could provide important and informative findings. It is possible that a particular sample design has strong influence on the quality of the methods and the results. Working with census data or register data (that are much larger than the Micro Census) lower variances and consequently lower MSE are expected and so the methods with small bias but high variance, like for example model (c)(ii) or model (e)(ii), would perform much better.

A. Appendix

A.1. R-Code

All the results are obtained with the statistical environment R [R Core Team, 2013].
Partly commands and functions from base R [R Core Team, 2013] and from the libraries
laeken [Alfons et al., 2013], sampling [Tillé and Matei, 2012], survey [Lumley, 2012, 2004],
robustbase [Rousseeuw et al., 2013, Todorov and Filzmoser, 2009], MASS [Venables and
Ripley, 2002], lme4 [Bates et al., 2013], VIM [Templ et al., 2013], StatMatch [D'Orazio,
2012], snow [Tierney et al., 2013] and rlecuyer [Sevcikova and Rossini, 2012] are used.

Special thanks are due to Alexander Kowarik, the cooperation, particularly concerning
the R-Code, has been a great help.

Important functions for the sampling of the EU-SILC data set, denoted in the following
also by **S**, and the Micro Census data set, denoted in the following also by **MZ**, out of
U are **strata** and **getdata**.
Furthermore the symbols $S^{(*)}$ and $MZ^{(*)}$, respectively, signify a bootstrap sample of **S**
and **MZ**, respectively.

The two samples **S** and **MZ** get combined by creating a list of two with **MZ** as the
first entry and **S** as the second list entry. So later on the access to **MZ** or **S** is done via
the operator ...[[1]] or ...[[2]].
In Listing 1 stands the code used for the sampling of **S** and **MZ** out of **U** using the
respective sample designs.

Listing 1: Sampling of **S** and **MZ**.

```
samples <- function(eusilcP, smhv_2_2011, seednr=FALSE){
        ziehsilc <- function(eusilcP, smhv_2_2011, seednr=FALSE){
                if (seednr!=FALSE){
                        set.seed(seednr)
                }
                size_silc <- data.frame(table(smhv_2_2011$strat[smhv_2_
                        2011$silc==1]))$Freq
                str_silc <- strata(eusilcP[order(eusilcP$strat),],
                        stratanames="strat", size=size_silc, method=c("
                        srswor"))
                silcsim <- getdata(eusilcP, str_silc)
                silcsim
        }

        ziehmz <- function(eusilcP, smhv_2_2011, seednr=FALSE){
                if (seednr!=FALSE){
                        set.seed(seednr)
                }
```

```
            size_mz <- data.frame(table(smhv_2_2011$strat[smhv_2_
               2011$silc==0]))$Freq
            str_mz <- strata(eusilcP[order(eusilcP$strat),],
               stratanames="strat", size=size_mz, method=c("srswor"
               ))
            mzsim <- getdata(eusilcP, str_mz)
            mzsim
    }

    s <- list(mzsim = ziehmz(eusilcP, smhv_2_2011), silcsim =
       ziehsilc(eusilcP, smhv_2_2011))
    s
}
```

The function argument and input for the functions performed and explained below (see, e.g., Listing 2) is a so called samp. This is a list including two elements, the simulated Micro Census and the simulated EU-SILC data set (or a bootstrap sample of the simulated Micro Census and EU-SILC data set). For **MZ** and **S** generally speaking this will be a list obtained from the function samples (compare Listing 1), so the first list entry samp[[1]]) equates to **MZ** and the second entry samp[[2]]) to **S**.

A.1.1. Regression Models Including Selected Small Area Methods

For the estimation of the linear regression models it is worked with the function lm. This function gets the model via the formula syntax and it gives the possibility to incorporate weights via the argument weights. The underlying data set is passed to the function via the argument data.

To impute the values in **MZ** and **MZ**$^{(*)}$, respectively, the function predict helps a lot. It needs the result of lm as input, as well as newdata, what is defined as **MZ** and **MZ**$^{(*)}$, respectively, in this case.

Out of the predicted and with the exponential function exp retransformed values get calculated the at-risk-of-poverty rate with the function arpr. This function needs the equivalised household income as input and it gives also the possibility to incorporate weights for the computation (argument weights). Furthermore it computes the at-risk-of-poverty rate for Austria and also separately for every state by defining the argument breakdown as strat.

In Listing 2 can be found the code for the estimation and transmission of the weighted OLS regression model with only socio-demographic variables and the estimation of the at-risk-of-poverty rate.

In Listing 3 stands the corresponding code using both, socio-demographic variables and some covariates, as predictor variables.

Listing 2: Ordinary Least Squares Regression: model estimation on the basis of EU-SILC (**S** or **S**$^{(*)}$ equates to samp[[2]]), transmission to Micro Census (**MZ** or **MZ**$^{(*)}$ equates to samp[[1]]) and calculation of arpr out of the imputed values: with socio-demographic variables.

```
gewlmreg_soz <- function(samp){
    samp[[2]]$rp_age <- (as.ordered(samp[[2]]$rp_age))
    samp[[2]]$workint <- (as.ordered(samp[[2]]$workint))
    samp[[2]]$rp_heduc <- (as.ordered(samp[[2]]$rp_heduc))
    samp[[2]]$equipm <- (as.ordered(samp[[2]]$equipm))
    samp[[1]]$rp_age <- (as.ordered(samp[[1]]$rp_age))
    samp[[1]]$workint <- (as.ordered(samp[[1]]$workint))
    samp[[1]]$rp_heduc <- (as.ordered(samp[[1]]$rp_heduc))
    samp[[1]]$equipm <- (as.ordered(samp[[1]]$equipm))
    var0 <- c("rp_age", "dwell", "htyp", "migration", "equipm", "rp
        _living", "rp_famst", "workint", "rp_heduc", "rp_occstat", "
        rp_branch", "urb")
    # in EQ_INC some 0 occur --> set to 1
    # (to make it possible to use log(EQ_INC))
    samp[[2]]$EQ_INC[samp[[2]]$EQ_INC==0] <- 1

    # estimate the model
    form0 <- as.formula(paste("log(EQ_INC)~", paste(var0,collapse="
        +",sep=""), sep=""))
    mod0 <- lm(form0, data=samp[[2]], weights=samp[[2]]$gew)
    # transmit the model to MZ and retransform
    samp[[1]]$EQ_INCe_0 <- exp(predict(mod0, newdata=samp[[1]]))

    # estimate the at-risk-of-poverty rate for Austria + 9 states
    ARPR_0 <- c(arpr("EQ_INCe_0", weights = "gew", breakdown = "
        strat", data = samp[[1]])$value, arpr("EQ_INCe_0", weights =
        "gew", breakdown = "strat", data = samp[[1]])$
        valueByStratum$value) /100
    ARPR_0
}
```

Listing 3: Ordinary Least Squares Regression: model estimation on the basis of EU-SILC (**S** or **S**$^{(*)}$ equates to samp[[2]]), transmission to Micro Census (**MZ** or **MZ**$^{(*)}$ equates to samp[[1]]) and calculation of arpr out of the imputed values: with socio-demographic variables + covariates.

```
gewlmreg_rV <- function(samp){
    samp[[2]]$rp_age <- (as.ordered(samp[[2]]$rp_age))
    samp[[2]]$workint <- (as.ordered(samp[[2]]$workint))
    samp[[2]]$rp_heduc <- (as.ordered(samp[[2]]$rp_heduc))
    samp[[2]]$equipm <- (as.ordered(samp[[2]]$equipm))
    samp[[1]]$rp_age <- (as.ordered(samp[[1]]$rp_age))
    samp[[1]]$workint <- (as.ordered(samp[[1]]$workint))
    samp[[1]]$rp_heduc <- (as.ordered(samp[[1]]$rp_heduc))
    samp[[1]]$equipm <- (as.ordered(samp[[1]]$equipm))
    var0 <- c("rp_age", "dwell", "htyp", "migration", "equipm", "rp
        _living", "rp_famst", "workint", "rp_heduc", "rp_occstat", "
```

```
      rp_branch", "urb")
    var1 <- c(var0, "migrbal", "nonempm", "nonempw", "quinc", "
      carscc", "unemp", "compall")
    # in EQ_INC some 0 occur --> set to 1
    # (to make it possible to use log(EQ_INC))
    samp[[2]]$EQ_INC[samp[[2]]$EQ_INC==0] <- 1

    # estimate the model
    form1 <- as.formula(paste("log(EQ_INC)~", paste(var1,collapse="
      +",sep=""), sep=""))
    mod1 <- lm(form1, data=samp[[2]], weights=samp[[2]]$gew)
    # transmit the model to MZ and retransform
    samp[[1]]$EQ_INCe_1 <- exp(predict(mod1, newdata=samp[[1]]))

    # estimate the at-risk-of-poverty rate for Austria + 9 states
    ARPR_1 <- c(arpr("EQ_INCe_1", weights = "gew", breakdown = "
      strat", data = samp[[1]])$value, arpr("EQ_INCe_1", weights =
      "gew", breakdown = "strat", data = samp[[1]])$
      valueByStratum$value) /100
    ARPR_1
}
```

Concerning the robust linear regression methods it is worked with the function rlm out of the package MASS, see Listing 4 and Listing 5. It is to handle in a similar way like the function lm concerning the formula syntax, the data and weights arguments. By defining the method as ''M'' it is possible to incorporate weights. Using this method a linear model is fitted by robust regression via Hubers M-estimator. It is the default for this function. Another possibility would be to work with method=''MM'', but weights are not supported for this version, so it isn't used for estimation.

The function predict works in connection with rlm, so it is used and the next steps are the same as for the ordinary linear regression method (function exp for the retransformation and the function arpr for the calculation of the at-risk-of-poverty rate).

See Listing 4 and Listing 5, respectively, for the code of the estimation and transmission of the robust linear regression models using once only socio-demographic variables and once socio-demographic variables and covariates, respectively, and the accordingly estimation of the at-risk-of-poverty rates of Austria and the nine regions.

Listing 4: Robust Linear Regression: model estimation on the basis of EU-SILC (**S** or **S**[(*)] equates to samp[[2]]), transmission to Micro Census (**MZ** or **MZ**[(*)] equates to samp[[1]]) and calculation of arpr out of the imputed values: with socio-demographic variables.

```
gewlmreg_sozrob <- function(samp){
    samp[[2]]$rp_age <- (as.ordered(samp[[2]]$rp_age))
    samp[[2]]$workint <- (as.ordered(samp[[2]]$workint))
    samp[[2]]$rp_heduc <- (as.ordered(samp[[2]]$rp_heduc))
    samp[[2]]$equipm <- (as.ordered(samp[[2]]$equipm))
    samp[[1]]$rp_age <- (as.ordered(samp[[1]]$rp_age))
```

```
samp[[1]]$workint <- (as.ordered(samp[[1]]$workint))
samp[[1]]$rp_heduc <- (as.ordered(samp[[1]]$rp_heduc))
samp[[1]]$equipm <- (as.ordered(samp[[1]]$equipm))
var0 <- c("rp_age", "dwell", "htyp", "migration", "equipm", "rp
    _living", "rp_famst", "workint", "rp_heduc", "rp_occstat", "
    rp_branch", "urb")
# in EQ_INC some 0 occur --> set to 1
# (to make it possible to use log(EQ_INC))
samp[[2]]$EQ_INC[samp[[2]]$EQ_INC==0] <- 1

# estimate the model
form0 <- as.formula(paste("log(EQ_INC)~", paste(var0,collapse="
    +",sep=""), sep=""))
mod2 <- rlm(form0, data=samp[[2]], weights=samp[[2]]$gew)
# transmit the model to MZ and retransform
samp[[1]]$EQ_INCe_r0 <- exp(predict(mod2, newdata=samp[[1]]))

# estimate the at-risk-of-poverty rate for Austria + 9 states
ARPR_r0 <- c(arpr("EQ_INCe_r0", weights = "gew", breakdown = "
    strat", data = samp[[1]])$value, arpr("EQ_INCe_r0", weights
    = "gew", breakdown = "strat", data = samp[[1]])$
    valueByStratum$value) /100
ARPR_r0
}
```

Listing 5: Robust Linear Regression: model estimation on the basis of EU-SILC ($\mathbf{S}$ or $\mathbf{S}^{(*)}$ equates to samp[[2]]), transmission to Micro Census ($\mathbf{MZ}$ or $\mathbf{MZ}^{(*)}$ equates to samp[[1]]) and calculation of arpr out of the imputed values: with sociodemographic variables + covariates.

```
gewlmreg_rVrob <- function(samp){
    samp[[2]]$rp_age <- (as.ordered(samp[[2]]$rp_age))
    samp[[2]]$workint <- (as.ordered(samp[[2]]$workint))
    samp[[2]]$rp_heduc <- (as.ordered(samp[[2]]$rp_heduc))
    samp[[2]]$equipm <- (as.ordered(samp[[2]]$equipm))
    samp[[1]]$rp_age <- (as.ordered(samp[[1]]$rp_age))
    samp[[1]]$workint <- (as.ordered(samp[[1]]$workint))
    samp[[1]]$rp_heduc <- (as.ordered(samp[[1]]$rp_heduc))
    samp[[1]]$equipm <- (as.ordered(samp[[1]]$equipm))
    var0 <- c("rp_age", "dwell", "htyp", "migration", "equipm", "rp
        _living", "rp_famst", "workint", "rp_heduc", "rp_occstat", "
        rp_branch", "urb")
    var2 <- c(var0, "migrbal", "nonempm", "nonempw", "quinc", "
        carscc", "unemp", "compall")
    # in EQ_INC some 0 occur --> set to 1
    # (to make it possible to use log(EQ_INC))
    samp[[2]]$EQ_INC[samp[[2]]$EQ_INC==0] <- 1

    # estimate the model
    form2 <- as.formula(paste("log(EQ_INC)~", paste(var2,collapse="
        +",sep=""), sep=""))
    mod3 <- rlm(form2, data=samp[[2]], weights=samp[[2]]$gew)
```

```
            # transmit the model to MZ and retransform
            samp[[1]]$EQ_INCe_r2 <- exp(predict(mod3, newdata=samp[[1]]))

            # estimate the at-risk-of-poverty rate for Austria + 9 states
            ARPR_r2 <- c(arpr("EQ_INCe_r2", weights = "gew", breakdown = "
                strat", data = samp[[1]])$value, arpr("EQ_INCe_r2", weights
                = "gew", breakdown = "strat", data = samp[[1]])$
                valueByStratum$value) /100
            ARPR_r2
}
```

Regarding the logistic regression, thus fitting a model for a binary response, it is worked
with the function glm (see Listing 6 and Listing 7). The argument family is defined as
quasibinomial, as described in the R help [compare R Core Team, 2013] of the function
glm and of family, it is the same as the binomial family, just "the dispersion parameter
is not fixed at one", so it is possible to "model over-dispersion". The binomial family corresponds to the above described logistic regression, meaning to the logit-transformation
(compare Chapter 2.3.3). For further details have a look in Sachs and Hedderich [2009].
It is again possible to incorporate weights with the function glm. They are passed to the
argument weights of the function.

The function predict is also implemented for a glm-object, so it is used to estimate the
values for the simulated Micro Census data sets. If a glm-object is passed to this function,
it is necessary to define the argument type. Because of the choice of type="response",
the predicted probabilities are retrieved. To estimate one of the two parameter-values
at-risk-of-poverty or not at-risk-of-poverty, samples according to these probabilities are
drawn, see Listing 6 and Listing 7.

To estimate the at-risk-of-poverty rate out of this dichotomous variable at-risk-of-
poverty a weighted mean is calculated with the function weighted.mean. On the one
hand it is calculated for all values of one simulated **MZ** or **MZ**[(*)], and on the other hand
it is calculated for subsets of them, namely for every of the nine states, see again Listing
6 and Listing 7.

The complete code concerning the logistic regression models (including the estimation
and transmission of the model and the calculation of the at-risk-of-poverty rates with the
estimated values) is written in Listing 6 and Listing 7. The difference of the two Listings
is in the choice of the predictor variables; once again only socio-demographic variables
are used, the other time is worked with socio-demographic variables and some covariates.

Listing 6: Logistic Regression: model estimation on the basis of EU-SILC (**S** or **S**[(*)]
        equates to samp[[2]]), transmission to Micro Census (**MZ** or **MZ**[(*)] equates
        to samp[[1]]) and calculation of arpr out of the imputed values: with socio-
        demographic variables.

```
logreg_soz <- function(samp){
        samp[[2]]$rp_age <- (as.ordered(samp[[2]]$rp_age))
```

```
samp[[2]]$workint <- (as.ordered(samp[[2]]$workint))
samp[[2]]$rp_heduc <- (as.ordered(samp[[2]]$rp_heduc))
samp[[2]]$equipm <- (as.ordered(samp[[2]]$equipm))
samp[[1]]$rp_age <- (as.ordered(samp[[1]]$rp_age))
samp[[1]]$workint <- (as.ordered(samp[[1]]$workint))
samp[[1]]$rp_heduc <- (as.ordered(samp[[1]]$rp_heduc))
samp[[1]]$equipm <- (as.ordered(samp[[1]]$equipm))
var0 <- c("rp_age", "dwell", "htyp", "migration", "equipm", "rp
    _living", "rp_famst", "workint", "rp_heduc", "rp_occstat", "
    rp_branch", "urb")

# estimate the model  (ARPT60i is a factor)
form0 <- as.formula(paste("ARPT60i~", paste(var0,collapse="+",
    sep=""), sep=""))
mod0 <- glm(form0, family=quasibinomial, data=samp[[2]],
    weights=samp[[2]]$gew)

# transmit the model to MZ
# in ARPT_0p: probabilities for at-risk-of-poverty = 1
# in ARPT_0: 0 or 1
samp[[1]]$ARPT_0p <- predict(mod0, type="response", newdata=
    samp[[1]])
samp[[1]]$ARPT_0 <- unlist(lapply(samp[[1]]$ARPT_0p, function(x
    ) sample(c(0,1),size=1,prob=c(1-x,x))))

# estimate the at-risk-of-poverty rate for Austria + 9 states
ARPR_0 <- weighted.mean(samp[[1]][,"ARPT_0"], samp[[1]]$gew)
for (b in 1:9){
        bb <- levels(samp[[1]]$strat)[b]   # state name
        # limit the data set and weight vector with samp[[1]]$
            strat==bb
        ARPR_0 <- c(ARPR_0, weighted.mean(samp[[1]][samp[[1]]$
            strat==bb,"ARPT_0"], samp[[1]]$gew[samp[[1]]$strat==
            bb]))
}
ARPR_0
}
```

Listing 7: Logistic Regression: model estimation on the basis of EU-SILC ($\mathbf{S}$ or $\mathbf{S}^{(*)}$ equates to samp[[2]]), transmission to Micro Census ($\mathbf{MZ}$ or $\mathbf{MZ}^{(*)}$ equates to samp[[1]]) and calculation of arpr out of the imputed values: with socio-demographic variables + covariates.

```
logreg_rV <- function(samp){
        samp[[2]]$rp_age <- (as.ordered(samp[[2]]$rp_age))
        samp[[2]]$workint <- (as.ordered(samp[[2]]$workint))
        samp[[2]]$rp_heduc <- (as.ordered(samp[[2]]$rp_heduc))
        samp[[2]]$equipm <- (as.ordered(samp[[2]]$equipm))
        samp[[1]]$rp_age <- (as.ordered(samp[[1]]$rp_age))
        samp[[1]]$workint <- (as.ordered(samp[[1]]$workint))
        samp[[1]]$rp_heduc <- (as.ordered(samp[[1]]$rp_heduc))
        samp[[1]]$equipm <- (as.ordered(samp[[1]]$equipm))
```

```
var0 <- c("rp_age", "dwell", "htyp", "migration", "equipm", "rp
    _living", "rp_famst", "workint", "rp_heduc", "rp_occstat", "
    rp_branch", "urb")
var1 <- c(var0, "migrbal", "nonempm", "nonempw", "quinc", "
    carscc", "unemp", "compall")

# estimate the model  (ARPT60i is a factor)
form1 <- as.formula(paste("ARPT60i~", paste(var1,collapse="+",
    sep=""), sep=""))
mod1 <- glm(form1, family=quasibinomial, data=samp[[2]],
    weights=samp[[2]]$gew)

# transmit the model to MZ
# in ARPT_0p: probabilities for at-risk-of-poverty = 1
# in ARPT_0: 0 or 1
samp[[1]]$ARPT_1p <- predict(mod1, type="response", newdata=
    samp[[1]])
samp[[1]]$ARPT_1 <- unlist(lapply(samp[[1]]$ARPT_1p, function(x
    ) sample(c(0,1),size=1,prob=c(1-x,x))))

# estimate the at-risk-of-poverty rate for Austria + 9 states
ARPR_1 <- weighted.mean(samp[[1]][,"ARPT_1"], samp[[1]]$gew)
for (b in 1:9){
        bb <- levels(samp[[1]]$strat)[b]    # state name
        # limit the data set and weight vector with samp[[1]]$
            strat==bb
        ARPR_1 <- c(ARPR_1, weighted.mean(samp[[1]][samp[[1]]$
            strat==bb,"ARPT_1"], samp[[1]]$gew[samp[[1]]$strat==
            bb]))
}
ARPR_1
}
```

In context of linear mixed models it is worked with the function lmer out of the package lme4 (see Listing 8 and Listing 9). This function expects as input a formula where on the left side of the sign $\sim$ the response stands and on the right side the predictor variables. As can be read in the R help of this function, compare Bates et al. [2013], "The vertical bar character | separates an expression for a model matrix and a grouping factor." Furthermore the data is passed to the function via the argument data and the weight vector via the argument weights as it is common for this type of functions. Generally and especially in the above introduced models, the weight for those observations not appearing in a bootstrap sample is set to zero (compare further down in Listing 24). So they are excluded in the model estimation, because in every method the weights are incorporated in the calculation. In connection with the function lmer this is not possible, because there must not be zeros in the weight vector. So the data set has to be restricted before by removing the non appearing observations, i.e. observations with weight 0, see Listing 8 and Listing 9. The output is an object of class mer.

The function `predict` is not implemented for objects of class `mer`, so it can't be used as done in the above introduced models. As a consequence, the prediction is done by hand. For this purpose information on the model is extracted in form of the "terms" object for the fixed-effects terms in the model formula with the command `terms`, see Listing 8 and Listing 9. With the function `model.matrix` a model matrix is created based on the information received from `terms` and the data, in this case the Micro Census data set. The command `fixef` applied to the model (i.e. to the output from `lmer`) returns the estimations of the parameters of the fixed effects. So the predictions are obtained via the matrix multiplication of the model matrix with the output of `fixef`. The result has to be retransformed again with the exponential function. (See Listing 8 and Listing 9.)

The at-risk-of-poverty rate gets calculated as usual with the function `arpr` (in connection with linear mixed models again the equivalised household income is estimated).

For the entire code used in context of linear mixed effects regression models, i.e. the model estimation, the transmission of the model and the estimation of the at-risk-of-poverty rates for Austria and the nine regions, have a look at Listing 8 and Listing 9, respectively. Listing 8 contains the version of the model where socio-demographic variables are used as fixed effects and state as random effect and Listing 9 contains the alternative with socio-demographic variables and some covariates as fixed effects and state as random effect.

Listing 8: Linear Mixed Regression: model estimation on the basis of EU-SILC ($\mathbf{S}$ or $\mathbf{S}^{(*)}$ equates to samp[[2]]), transmission to Micro Census ($\mathbf{MZ}$ or $\mathbf{MZ}^{(*)}$ equates to samp[[1]]) and calculation of arpr out of the imputed values: with socio-demographic variables as fixed effects and state as random effect.

```
mixed_lm_inc_soz <- function(samp){
    samp[[2]]$rp_age <- (as.ordered(samp[[2]]$rp_age))
    samp[[2]]$workint <- (as.ordered(samp[[2]]$workint))
    samp[[2]]$rp_heduc <- (as.ordered(samp[[2]]$rp_heduc))
    samp[[2]]$equipm <- (as.ordered(samp[[2]]$equipm))
    samp[[1]]$rp_age <- (as.ordered(samp[[1]]$rp_age))
    samp[[1]]$workint <- (as.ordered(samp[[1]]$workint))
    samp[[1]]$rp_heduc <- (as.ordered(samp[[1]]$rp_heduc))
    samp[[1]]$equipm <- (as.ordered(samp[[1]]$equipm))
    var0 <- c("rp_age", "dwell", "htyp", "migration", "equipm", "
        rp_living", "rp_famst", "workint", "rp_heduc", "rp_occstat",
        "rp_branch", "urb")
    # in EQ_INC some 0 occur --> set to 1
    # (to make it possible to use log(EQ_INC))
    samp[[2]]$EQ_INC[samp[[2]]$EQ_INC==0] <- 1

    # restriction of the data set
    # because there must not be zeros in the weight vector in lmer
    samp[[2]] <- samp[[2]][samp[[2]]$gew!=0,]
    samp[[1]] <- samp[[1]][samp[[1]]$gew!=0,]

    # estimate the model
    form0 <- as.formula(paste("log(EQ_INC)~", paste(var0,collapse="
```

73

```
                    +",sep=""), "+ (1|strat)", sep=""))
        mod0 <- lmer(form0, data=samp[[2]], weights=samp[[2]]$gew)
        # transmit the model to MZ and retransform
        samp[[1]]$EQ_INCe_0 <- exp(model.matrix(terms(mod0),samp[[1]])
            %*% fixef(mod0))

        # estimate the at-risk-of-poverty rate for Austria + 9 states
        ARPR_0 <- c(arpr("EQ_INCe_0", weights = "gew", breakdown = "
            strat", data = samp[[1]])$value, arpr("EQ_INCe_0", weights =
            "gew", breakdown = "strat", data = samp[[1]])$
            valueByStratum$value) /100
        ARPR_0
}
```

Listing 9: Linear Mixed Regression: model estimation on the basis of EU-SILC (**S** or **S**[(*)] equates to samp[[2]]), transmission to Micro Census (**MZ** or **MZ**[(*)] equates to samp[[1]]) and calculation of arpr out of the imputed values: with sociodemographic variables + covariates as fixed effects and state as random effect.

```
mixed_lm_inc_rV <- function(samp){
        samp[[2]]$rp_age <- (as.ordered(samp[[2]]$rp_age))
        samp[[2]]$workint <- (as.ordered(samp[[2]]$workint))
        samp[[2]]$rp_heduc <- (as.ordered(samp[[2]]$rp_heduc))
        samp[[2]]$equipm <- (as.ordered(samp[[2]]$equipm))
        samp[[1]]$rp_age <- (as.ordered(samp[[1]]$rp_age))
        samp[[1]]$workint <- (as.ordered(samp[[1]]$workint))
        samp[[1]]$rp_heduc <- (as.ordered(samp[[1]]$rp_heduc))
        samp[[1]]$equipm <- (as.ordered(samp[[1]]$equipm))
        var0 <- c("rp_age", "dwell", "htyp", "migration", "equipm", "rp
            _living", "rp_famst", "workint", "rp_heduc", "rp_occstat", "
            rp_branch", "urb")
        var1 <- c(var0, "migrbal", "nonempm", "nonempw", "quinc", "
            carscc", "unemp", "compall")
        # in EQ_INC some 0 occur --> set to 1
        # (to make it possible to use log(EQ_INC))
        samp[[2]]$EQ_INC[samp[[2]]$EQ_INC==0] <- 1

        # restriction of the data set
        # because there must not be zeros in the weight vector in lmer
        samp[[2]] <- samp[[2]][samp[[2]]$gew!=0,]
        samp[[1]] <- samp[[1]][samp[[1]]$gew!=0,]

        # estimate the model
        form1 <- as.formula(paste("log(EQ_INC)~", paste(var1,collapse="
            +",sep=""), "+ (1|strat)", sep=""))
        mod1 <- lmer(form1, data=samp[[2]], weights=samp[[2]]$gew)
        # transmit the model to MZ and retransform
        samp[[1]]$EQ_INCe_1 <- exp(model.matrix(terms(mod1),samp[[1]])
            %*% fixef(mod1))

        # estimate the at-risk-of-poverty rate for Austria + 9 states
        ARPR_1 <- c(arpr("EQ_INCe_1", weights = "gew", breakdown = "
```

```
        strat", data = samp[[i]])$value, arpr("EQ_INCe_1", weights =
        "gew", breakdown = "strat", data = samp[[i]])$
        valueByStratum$value) /100
    ARPR_1
}
```

Concerning logistic mixed regression models the used important function is `glmer` out of the package `lme4` [see Bates et al., 2013], see Listing 10 and Listing 11, respectively. This function works in the same way as `lmer`, the used function in context of linear mixed regression. The only difference is, that a `family` has to be defined. The `family` is defined as `binomial`, what comes up to the above described method (see Chapter 2.3.5). Furthermore again a formula is passed to the function as input, the data is alloted to the function via the argument `data`, the weight vector via the argument `weights` and there must not be zeros in the weight vector. So the data set gets restricted before by removing the non appearing observations, i.e. observations with weight 0, see Listing 10 and Listing 11, respectively. The output is again an object of class `mer`.

As mentioned above, the function `predict` is not implemented for objects of class `mer`, so it can't be used. So afresh the prediction is done by hand. For this purpose the commands `terms`, `model.matrix` and `fixef` are used one-to-one as for the linear mixed models (see Listing 10 and Listing 11), but contrary to linear mixed regression, the matrix multiplication of the model matrix with the output of `fixef` clearly doesn't yield the definite predictions. The result is in fact the prediction associated with the linear model as described in Chapter 2.3.5 (i.e. logit of the probabilities). The function `plogis` is used to retransform it and so to get the predicted probabilities (`plogis` is called the "inverse logit"). Again the probabilities are discretised by drawing samples with these probabilities in order to get one of the two parameter values at-risk-of-poverty or not at-risk-of-poverty as prediction of the dichotomous factor, see Listing 10 and Listing 11, respectively.

To estimate the at-risk-of-poverty rate out of these values of the dichotomous variable at-risk-of-poverty a weighted mean is calculated again with the function `weighted.mean`, on the one hand for all predicted values of one simulated **MZ** or **MZ**$^{(*)}$ and on the other hand for subsets of them, namely for every of the nine states.

In Listing 10 can be found the code for the estimation and transmission of the logistic mixed regression model with only socio-demographic variables as fixed effects (and state as random effect) and the estimation of the at-risk-of-poverty rate.

In Listing 11 stands the corresponding code using both, socio-demographic variables and some covariates, as fixed effects (and state as random effect).

Listing 10: Logistic Mixed Regression: model estimation on the basis of EU-SILC (**S** or **S**[(*)] equates to samp[[2]]), transmission to Micro Census (**MZ** or **MZ**[(*)] equates to samp[[1]]) and calculation of arpr out of the imputed values: with socio-demographic variables as fixed effects and state as random effect.

```
mixed_logm_arpt_soz <- function(samp){
        samp[[2]]$rp_age <- (as.ordered(samp[[2]]$rp_age))
        samp[[2]]$workint <- (as.ordered(samp[[2]]$workint))
        samp[[2]]$rp_heduc <- (as.ordered(samp[[2]]$rp_heduc))
        samp[[2]]$equipm <- (as.ordered(samp[[2]]$equipm))
        samp[[1]]$rp_age <- (as.ordered(samp[[1]]$rp_age))
        samp[[1]]$workint <- (as.ordered(samp[[1]]$workint))
        samp[[1]]$rp_heduc <- (as.ordered(samp[[1]]$rp_heduc))
        samp[[1]]$equipm <- (as.ordered(samp[[1]]$equipm))
        var0 <- c("rp_age", "dwell", "htyp", "migration", "equipm", "
            rp_living", "rp_famst", "workint", "rp_heduc", "rp_occstat",
            "rp_branch", "urb")

        # restriction of the data set
        # because there must not be zeros in the weight vector in lmer
        samp[[2]] <- samp[[2]][samp[[2]]$gew!=0,]
        samp[[1]] <- samp[[1]][samp[[1]]$gew!=0,]

        # estimate the model  (ARPT60i is a factor)
        form0 <- as.formula(paste("ARPT60i~", paste(var0,collapse="+",
            sep=""), "+ (1|strat)", sep=""))
        mod0 <- glmer(form0, family=binomial, data=samp[[2]], weights=
            samp[[2]]$gew)
        # transmit the model to MZ
        # in ARPT_0lo: logit of the probabilities for at-risk-of-
            poverty
        # in ARPT_0p: probabilities for at-risk-of-poverty = 1
        # in ARPT_0: 0 or 1
        ARPT_0lo <- model.matrix(terms(mod0),samp[[1]]) %*% fixef(mod0)
        samp[[1]]$ARPT_0p <- plogis(ARPT_0lo)
        samp[[1]]$ARPT_0 <- unlist(lapply(samp[[1]]$ARPT_0p, function(x
            )sample(c(0,1),size=1,prob=c(1-x,x))))

        # estimate the at-risk-of-poverty rate for Austria + 9 states
        ARPR_0 <- weighted.mean(samp[[1]][,"ARPT_0"], samp[[1]]$gew)
        for (b in 1:9){
                bb <- levels(samp[[1]]$strat)[b]    # state name
                # limit the data set and weight vector with samp[[1]]$
                    strat==bb
                ARPR_0 <- c(ARPR_0, weighted.mean(samp[[1]][samp[[1]]$
                    strat==bb,"ARPT_0"],samp[[1]]$gew[samp[[1]]$strat==
                    bb]))
        }
        ARPR_0
}
```

Listing 11: Logistic Mixed Regression: model estimation on the basis of EU-SILC (**S** or **S**$^{(*)}$ equates to **samp[[2]]**), transmission to Micro Census (**MZ** or **MZ**$^{(*)}$ equates to **samp[[1]]**) and calculation of arpr out of the imputed values: with socio-demographic variables + covariates as fixed effects and state as random effect.

```
mixed_logm_arpt_rV <- function(samp){
    samp[[2]]$rp_age <- (as.ordered(samp[[2]]$rp_age))
    samp[[2]]$workint <- (as.ordered(samp[[2]]$workint))
    samp[[2]]$rp_heduc <- (as.ordered(samp[[2]]$rp_heduc))
    samp[[2]]$equipm <- (as.ordered(samp[[2]]$equipm))
    samp[[1]]$rp_age <- (as.ordered(samp[[1]]$rp_age))
    samp[[1]]$workint <- (as.ordered(samp[[1]]$workint))
    samp[[1]]$rp_heduc <- (as.ordered(samp[[1]]$rp_heduc))
    samp[[1]]$equipm <- (as.ordered(samp[[1]]$equipm))
    var0 <- c("rp_age", "dwell", "htyp", "migration", "equipm", "
        rp_living", "rp_famst", "workint", "rp_heduc", "rp_occstat",
        "rp_branch", "urb")
    var1 <- c(var0, "migrbal", "nonempm", "nonempw", "quinc", "
        carscc", "unemp", "compall")

    # restriction of the data set
    # because there must not be zeros in the weight vector in lmer
    samp[[2]] <- samp[[2]][samp[[2]]$gew!=0,]
    samp[[1]] <- samp[[1]][samp[[1]]$gew!=0,]

    # estimate the model (ARPT60i is a factor)
    form1 <- as.formula(paste("ARPT60i~", paste(var1,collapse="+",
        sep=""), "+ (1|strat)", sep=""))
    mod1 <- glmer(form1, family=binomial, data=samp[[2]], weights=
        samp[[2]]$gew)
    # transmit the model to MZ
    # in ARPT_1lo: logit of the probabilities for at-risk-of-
        poverty
    # in ARPT_1p: probabilities for at-risk-of-poverty = 1
    # in ARPT_1: 0 or 1
    ARPT_1lo <- model.matrix(terms(mod1),samp[[1]]) %*% fixef(mod1)
    samp[[1]]$ARPT_1p <- plogis(ARPT_1lo)
    samp[[1]]$ARPT_1 <- unlist(lapply(samp[[1]]$ARPT_1p, function(x
        )sample(c(0,1),size=1,prob=c(1-x,x))))

    # estimate the at-risk-of-poverty rate for Austria + 9 states
    ARPR_1 <- weighted.mean(samp[[1]][,"ARPT_1"], samp[[1]]$gew)
    for (b in 1:9){
            bb <- levels(samp[[1]]$strat)[b]  # state name
            # limit the data set and weight vector with samp[[1]]$
                strat==bb
            ARPR_1 <- c(ARPR_1, weighted.mean(samp[[1]][samp[[1]]$
                strat==bb,"ARPT_1"],samp[[1]]$gew[samp[[1]]$strat==
                bb]))
    }
    ARPR_1
}
```

77

A.1.2. Statistical Matching

In connection with statistical matching it is mainly worked with two different packages, namely with the packages VIM [see Templ et al., 2013] and StatMatch [see D'Orazio, 2012].

As can be read above there are implemented three different methods in the context of statistical matching. For two of them, namely random hot deck and sequential random hot deck, it is worked with a function out of the package VIM and for the last one, namely weighted random hot deck, functions out of the package StatMatch are used.

The main used function to implement random hot deck is hotdeck out of the package VIM [see Templ et al., 2013]. For this purpose the data sets are edited in the sense that one data set is constructed containing both, the Micro Census and the EU-SILC data set, whereas the values for ARPT60i and EQ_INC are missing in the Micro Census data set. This single data set is passed to the function hotdeck in the first place, see Listing 12, Listing 13, Listing 14 and Listing 15. Via the argument variable the variable, where missing values should be imputed, is defined. Furthermore the names of the variables, that should be used for the building of donation classes, are passed to the argument domain_var. Within these domains the imputation happens. The output of the function is the previously passed data set, but now with the imputed values.

In the case of the imputation of the factor variable ARPT60i again with the function weighted.mean the at-risk-of-poverty rate is calculated out of this dichotomous variable at-risk-of-poverty. On the one hand it is computed for all imputed values of one simulated **MZ** or **MZ**[(*)], and on the other hand it is computed for subsets of them, namely for every of the nine states. (See Listing 12 and Listing 14.)

On the other side, in the case of the imputation of the numeric variable EQ_INC the calculation of the at-risk-of-poverty rates is done again out of the imputed values with the function arpr (see Listing 13 and Listing 15). As a quick reminder, this function needs the equivalised household income as input and via the argument weights the weights are incorporated for the computation. It computes the at-risk-of-poverty rate for Austria and also separately for every state by defining the argument breakdown as strat.

Listing 12 contains the R-Code for random hot deck imputation of ARPT60i using the states for building donation classes and the R-Code for the estimation of the at-risk-of-poverty rates.

Listing 13 is the analogue for the imputation of EQ_INC.

In Listing 14 (imputation of ARPT60i) and Listing 15 (imputation of EQ_INC) are the R-Codes of alternatives hereof using the variables state and foreign origin of the household for the building of donation classes.

Listing 12: Random Hot Deck: imputation of ARPT60i in the Micro Census (MZ or MZ$^{(*)}$ equates to samp[[1]]) with the usage of EU-SILC (S or S$^{(*)}$ equates to samp[[2]]) and calculation of arpr out of the imputed values: with the states as donation classes.

```
match_strat_arpt <- function(samp){
    samp[[1]]$ARPT60i <- NA samp[[1]]$EQ_INC <- NA
    # combine the data sets MZ and S (MZ* and S*)
    data <- rbind(samp[[1]],samp[[2]])[,c("rp_age", "dwell", "htyp"
    , "migration", "equipm", "rp_living", "rp_famst", "workint",
    "rp_heduc", "rp_occstat", "rp_branch", "EQ_INC", "ARPT60i",
    "urb", "migrbal", "nonempm", "nonempw", "birth", "mort", "
    quinc", "carscc", "unemp", "socass", "compall", "unemp_lag",
    "socass_lag", "compall_lag", "strat", "gew")]
    # exclude observations with weight 0
    data <- data[data$gew!=0,]

    group.v1 <- "strat"   # variable for building domains
    # imputation of ARPT60i within the domains:
    dataI.1 <- hotdeck(data, variable="ARPT60i", domain_var=
        group.v1)

    # estimate the at-risk-of-poverty rate for Austria + 9 states
    ARPR <- weighted.mean(as.numeric(dataI.1$ARPT60i[1:sum(samp
        [[1]]$gew>0)]), weights="gew")
    for (b in 1:9){
        bb <- levels(samp[[1]]$strat)[b]   # state name
        # limit the data set and weight vector with samp[[1]]$
            strat==bb
        ARPR <- c(ARPR, weighted.mean(as.numeric(dataI.1[1:sum(
            samp[[1]]$gew>0),][dataI.1[1:sum(samp[[1]]$gew>0),]$
            strat==bb,]$ARPT60i), weights="gew"))
    }
    ARPR
}
```

Listing 13: Random Hot Deck: imputation of EQ_INC in the Micro Census (MZ or MZ$^{(*)}$ equates to samp[[1]]) with the usage of EU-SILC (S or S$^{(*)}$ equates to samp[[2]]) and calculation of arpr out of the imputed values: with the states as donation classes.

```
match_strat_inc <- function(samp){
    samp[[1]]$ARPT60i <- NA
    samp[[1]]$EQ_INC <- NA
    # combine the data sets MZ and S (MZ* and S*)
    data <- rbind(samp[[1]],samp[[2]])[,c("rp_age", "dwell", "htyp"
    , "migration", "equipm", "rp_living", "rp_famst", "workint",
    "rp_heduc", "rp_occstat", "rp_branch", "EQ_INC", "ARPT60i",
    "urb", "migrbal", "nonempm", "nonempw", "birth", "mort", "
    quinc", "carscc", "unemp", "socass", "compall", "unemp_lag",
    "socass_lag", "compall_lag", "strat", "gew")]
    # exclude observations with weight 0
```

```
        data <- data[data$gew!=0,]

        group.v1 <- "strat"   # variable for building domains
        # imputation of EQ_INC within the domains:
        dataI.1 <- hotdeck(data, variable="EQ_INC", domain_var=group.v1
            )

        # estimate the at-risk-of-poverty rate for Austria + 9 states
        ARPR <- c(arpr("EQ_INC", weights = "gew", breakdown = "strat",
            data = dataI.1[1:sum(samp[[1]]$gew>0),])$value, arpr("EQ_INC
            ", weights = "gew", breakdown = "strat", data = dataI.1[1:
            sum(samp[[1]]$gew>0),])$valueByStratum$value) /100
        ARPR
}
```

Listing 14: Random Hot Deck: imputation of `ARPT60i` in the Micro Census (**MZ** or **MZ**[(*)] equates to samp[[1]]) with the usage of EU-SILC (**S** or **S**[(*)] equates to samp[[2]]) and calculation of arpr out of the imputed values: with the usage of the variables state and foreign origin of the household for the building of donation classes.

```
match_stratmigr_arpt <- function(samp){
        samp[[1]]$ARPT60i <- NA
        samp[[1]]$EQ_INC <- NA
        # combine the data sets MZ and S (MZ* and S*)
        data <- rbind(samp[[1]],samp[[2]])[,c("rp_age", "dwell", "htyp"
            , "migration", "equipm", "rp_living", "rp_famst", "workint",
            "rp_heduc", "rp_occstat", "rp_branch", "EQ_INC", "ARPT60i",
            "urb", "migrbal", "nonempm", "nonempw", "birth", "mort", "
            quinc", "carscc", "unemp", "socass", "compall", "unemp_lag",
            "socass_lag", "compall_lag", "strat", "gew")]
        # exclude observations with weight 0
        data <- data[data$gew!=0,]

        group.v2 <- c("strat", "migration")   # variables for building
            domains
        # imputation of ARPT60i within the domains:
        dataI.2 <- hotdeck(data, variable="ARPT60i", domain_var=
            group.v2)

        # estimate the at-risk-of-poverty rate for Austria + 9 states
        ARPR <- weighted.mean(as.numeric(dataI.2$ARPT60i[1:sum(samp
            [[1]]$gew>0)]), weights="gew")
        for (b in 1:9){
                bb <- levels(samp[[1]]$strat)[b]   # state name
                # limit the data set and weight vector with samp[[1]]$
                    strat==bb
                ARPR <- c(ARPR, weighted.mean(as.numeric(dataI.2[1:sum(
                    samp[[1]]$gew>0),][dataI.2[1:sum(samp[[1]]$gew>0),]$
                    strat==bb,]$ARPT60i), weights="gew"))
        }
```

Listing 15: Random Hot Deck: imputation of EQ_INC in the Micro Census (**MZ** or **MZ**[(*)] equates to samp[[1]]) with the usage of EU-SILC (**S** or **S**[(*)] equates to samp[[2]]) and calculation of arpr out of the imputed values: with the usage of the variables state and foreign origin of the household for the building of donation classes.

```
match_stratmigr_inc <- function(samp){
    samp[[1]]$ARPT60i <- NA
    samp[[1]]$EQ_INC <- NA
    # combine the data sets MZ and S (MZ* and S*)
    data <- rbind(samp[[1]],samp[[2]])[,c("rp_age", "dwell", "htyp"
        , "migration", "equipm", "rp_living", "rp_famst", "workint",
        "rp_heduc", "rp_occstat", "rp_branch", "EQ_INC", "ARPT60i",
        "urb", "migrbal", "nonempm", "nonempw", "birth", "mort", "
        quinc", "carscc", "unemp", "socass", "compall", "unemp_lag",
        "socass_lag", "compall_lag", "strat", "gew")]
    # exclude observations with weight 0
    data <- data[data$gew!=0,]

    group.v2 <- c("strat", "migration")   # variables for building
        domains
    # imputation of EQ_INC within the domains:
    dataI.2 <- hotdeck(data, variable="EQ_INC", domain_var=group.v2
        )

    # estimate the at-risk-of-poverty rate for Austria + 9 states
    ARPR <- c(arpr("EQ_INC", weights = "gew", breakdown = "strat",
        data = dataI.2[1:sum(samp[[1]]$gew>0),])$value, arpr("EQ_INC
        ", weights = "gew", breakdown = "strat", data = dataI.2[1:
        sum(samp[[1]]$gew>0),])$valueByStratum$value) /100
    ARPR
}
```

For sequential random hot deck methods it is worked again with the function hotdeck out of the package VIM. What makes the only difference in the code of the random hot deck models and the one of the sequential random hot deck models is the additional definition of the argument ord_var in the function hotdeck, see Listing 16, Listing 17, Listing 18 and Listing 19, respectively. This argument gets the names of the variables used for the sorting of the data set before the imputation step. As described above, the ordering happens within every donation class. Hence the building of one data set for the input of the function in the first place, the definition of variable and domain_var and the output constituted by the imputed data set pass off analogously to the random hot deck models.

Again the at-risk-of-poverty rate is calculated as weighted mean in the case of the imputation of the factor variable (see Listing 16 and Listing 18) or with the function arpr in the case of the imputation of the numeric variable (see Listing 17 and Listing 19) as can be seen already further up.

In Listing 16 and Listing 17, respectively, can be found the R-Code concerning the imputation of ARPT60i and EQ_INC, respectively, with sequential random hot deck methods using the variables state and foreign origin of the household for the building of donation classes and using the variables occupational status, work intensity of the household, highest completed level of education, living, age class and category of the equipment of the apartment as ordering variables and the R-Code for the estimation of the at-risk-of-poverty rates for Austria and the nine regions.

Listing 18 and Listing 19 contains basically the same, but with the usage of more or less all appearing variables as ordering variables now.

Listing 16: Sequential Random Hot Deck: imputation of ARPT60i in the Micro Census (**MZ** or **MZ**$^{(*)}$ equates to samp[[1]]) with the usage of EU-SILC (**S** or **S**$^{(*)}$ equates to samp[[2]]) and calculation of arpr out of the imputed values: with the usage of the variables state and foreign origin of the household for the building of donation classes and with the usage of the variables occupational status, work intensity of the household, highest completed level of education, living, age class and category of the equipment of the apartment as ordering variables.

```
match_stratmigr_arpt_feklaw <- function(samp){
    samp[[1]]$ARPT60i <- NA
    samp[[1]]$EQ_INC <- NA
    # combine the data sets MZ and S (MZ* and S*)
    data <- rbind(samp[[1]],samp[[2]])[,c("rp_age", "dwell", "htyp"
        , "migration", "equipm", "rp_living", "rp_famst", "workint",
        "rp_heduc", "rp_occstat", "rp_branch", "EQ_INC", "ARPT60i",
        "urb", "migrbal", "nonempm", "nonempw", "birth", "mort", "
        quinc", "carscc", "unemp", "socass", "compall", "unemp_lag",
        "socass_lag", "compall_lag", "strat", "gew")]
    # exclude observations with weight 0
    data <- data[data$gew!=0,]

    group.v2 <- c("strat", "migration")    # variables for building
        domains
    # imputation of ARPT60i within the domains using the listed
        ordering variables:
    dataI.2 <- hotdeck(data, variable="ARPT60i", domain_var=
        group.v2, ord_var=c("rp_occstat","workint","rp_heduc","rp_
        living","rp_age","equipm"))

    # estimate the at-risk-of-poverty rate for Austria + 9 states
    ARPR <- weighted.mean(as.numeric(dataI.2$ARPT60i[1:sum(samp
        [[1]]$gew>0)]), weights="gew")
    for (b in 1:9){
        bb <- levels(samp[[1]]$strat)[b]    # state name
            # limit the data set and weight vector with samp[[1]]$
                strat==bb
```

```
        ARPR <- c(ARPR, weighted.mean(as.numeric(dataI.2[1:sum(
        samp[[1]]$gew>0),][dataI.2[1:sum(samp[[1]]$gew>0),]$
        strat==bb,]$ARPT60i), weights="gew"))
    }
    ARPR
}
```

Listing 17: Sequential Random Hot Deck: imputation of EQ_INC in the Micro Census (**MZ** or **MZ**[*] equates to samp[[1]]) with the usage of EU-SILC (**S** or **S**[*] equates to samp[[2]]) and calculation of arpr out of the imputed values: with the usage of the variables state and foreign origin of the household for the building of donation classes and with the usage of the variables occupational status, work intensity of the household, highest completed level of education, living, age class and category of the equipment of the apartment as ordering variables.

```
match_stratmigr_inc_feklaw <- function(samp){
    samp[[1]]$ARPT60i <- NA
    samp[[1]]$EQ_INC <- NA
    # combine the data sets MZ and S (MZ* and S*)
    data <- rbind(samp[[1]],samp[[2]])[,c("rp_age", "dwell", "htyp"
        , "migration", "equipm", "rp_living", "rp_famst", "workint",
        "rp_heduc", "rp_occstat", "rp_branch", "EQ_INC", "ARPT60i",
        "urb", "migrbal", "nonempm", "nonempw", "birth", "mort", "
        quinc", "carscc", "unemp", "socass", "compall", "unemp_lag",
        "socass_lag", "compall_lag", "strat", "gew")]
    # exclude observations with weight 0
    data <- data[data$gew!=0,]

    group.v2 <- c("strat", "migration")    # variables for building
        domains
    # imputation of EQ_INC within the domains using the listed
        ordering variables:
    dataI.2 <- hotdeck(data, variable="EQ_INC", domain_var=group.v2
        , ord_var=c("rp_occstat","workint", "rp_heduc", "rp_living",
        "rp_age", "equipm"))

    # estimate the at-risk-of-poverty rate for Austria + 9 states
    ARPR <- c(arpr("EQ_INC", weights = "gew", breakdown = "strat",
        data = dataI.2[1:sum(samp[[1]]$gew>0),])$value, arpr("EQ_INC
        ", weights = "gew", breakdown = "strat", data = dataI.2[1:
        sum(samp[[1]]$gew>0),])$valueByStratum$value) /100
    ARPR
}
```

Listing 18: Sequential Random Hot Deck: imputation of ARPT60i in the Micro Census (**MZ** or **MZ**[(*)] equates to samp[[1]]) with the usage of EU-SILC (**S** or **S**[(*)] equates to samp[[2]]) and calculation of arpr out of the imputed values: with the usage of the variables state and foreign origin of the household for the building of donation classes and with the usage of more or less all appearing variables as ordering variables.

```
match_stratmigr_arpt_ALLE <- function(samp){
    samp[[1]]$ARPT60i <- NA
    samp[[1]]$EQ_INC <- NA
    # combine the data sets MZ and S (MZ* and S*)
    data <- rbind(samp[[1]],samp[[2]])[,c("rp_age", "dwell", "htyp"
        , "migration", "equipm", "rp_living", "rp_famst", "workint",
        "rp_heduc", "rp_occstat", "rp_branch", "EQ_INC", "ARPT60i",
        "urb", "migrbal", "nonempm", "nonempw", "birth", "mort", "
        quinc", "carscc", "unemp", "socass", "compall", "unemp_lag",
        "socass_lag", "compall_lag", "strat", "gew")]
    # exclude observations with weight 0
    data <- data[data$gew!=0,]

    group.v2 <- c("strat", "migration")    # variables for building
        domains
    # imputation of ARPT60i within the domains using the listed
        ordering variables:
    dataI.2 <- hotdeck(data, variable="ARPT60i", domain_var=
        group.v2, ord_var=names(data)[-c(which(names(data)=="EQ_INC"
        ), which(names(data)=="ARPT60i"), which(names(data)=="urb"),
        which(names(data)=="gew"))])

    # estimate the at-risk-of-poverty rate for Austria + 9 states
    ARPR <- weighted.mean(as.numeric(dataI.2$ARPT60i[1:sum(samp
        [[1]]$gew>0)]), weights="gew")
    for (b in 1:9){
            bb <- levels(samp[[1]]$strat)[b]    # state name
            # limit the data set and weight vector with samp[[1]]$
                strat==bb
            ARPR <- c(ARPR, weighted.mean(as.numeric(dataI.2[1:sum(
                samp[[1]]$gew>0),][dataI.2[1:sum(samp[[1]]$gew>0),]$
                strat==bb,]$ARPT60i), weights="gew"))
    }
    ARPR
}
```

Listing 19: Sequential Random Hot Deck: imputation of EQ_INC in the Micro Census (**MZ** or **MZ**[(*)] equates to samp[[1]]) with the usage of EU-SILC (**S** or **S**[(*)] equates to samp[[2]]) and calculation of arpr out of the imputed values: with the usage of the variables state and foreign origin of the household for the building of donation classes and with the usage of more or less all appearing variables as ordering variables.

```
match_stratmigr_inc_ALLE <- function(samp){
    samp[[1]]$ARPT60i <- NA
```

```
    samp[[1]]$EQ_INC <- NA
    # combine the data sets MZ and S (MZ* and S*)
    data <- rbind(samp[[1]],samp[[2]])[,c("rp_age", "dwell", "htyp"
    , "migration", "equipm", "rp_living", "rp_famst", "workint",
    "rp_heduc", "rp_occstat", "rp_branch", "EQ_INC", "ARPT60i",
    "urb", "migrbal", "nonempm", "nonempw", "birth", "mort", "
    quinc", "carscc", "unemp", "socass", "compall", "unemp_lag",
    "socass_lag", "compall_lag", "strat", "gew")]
    # exclude observations with weight 0
    data <- data[data$gew!=0,]

    group.v2 <- c("strat", "migration")   # variables for building
        domains
    # imputation of EQ_INC within the domains using the listed
        ordering variables:
    dataI.2 <- hotdeck(data, variable="EQ_INC", domain_var=group.v2
    , ord_var=names(data)[-c(which(names(data)=="EQ_INC"), which
    (names(data)=="ARPT60i"), which(names(data)=="urb"), which(
    names(data)=="gew"))])

    # estimate the at-risk-of-poverty rate for Austria + 9 states
    ARPR <- c(arpr("EQ_INC", weights = "gew", breakdown = "strat",
    data = dataI.2[1:sum(samp[[1]]$gew>0),])$value, arpr("EQ_INC
    ", weights = "gew", breakdown = "strat", data = dataI.2[1:
    sum(samp[[1]]$gew>0),])$valueByStratum$value) /100
    ARPR
}
```

In context of weighted random hot deck models it is mainly worked with two functions
out of the package StatMatch.

The first one is RANDwNND.hotdeck. This function requires as input the recipient data
set and the donor data set. They are passed to the arguments data.rec and data.don,
respectively. Furthermore the argument match.vars gets the names of the variables that
should be used for the computation of the distances between the observations of the donor
and the recipient data set. As done here, it can be defined as NULL, i.e. no matching
variables are considered, with the consequence that all the units in the same donation
class can be chosen as donors, see Listing 20, Listing 21, Listing 22 and Listing 23. The
selection of one donor passes off with probability proportional to its weight by defining
the argument weight.don with the name of the variable, that contains the weights of the
donors. The domains are built by defining the argument don.class with the names of
the variables that should be used therefor. To the arguments dist.fun, cut.don and k
could be passed the distance function that should be used, the method that should be used
to form the nearest donor oberservations and a number whose meaning depends on the
choice of cut.don. For example, the distance function is defined as Gower, indicating that
the Gower distance is used, and cut.don is defined as k.dist, indicating that possible
donors are those observations with distance from the recipient less or equal to the value
specified with the argument k. The output of this function is a list containing among

85

others a matrix indicating the row names/numbers of the units of the recipient data set with the corresponding row names/numbers of the selected donor units. This matrix is accessed via the list element `mtc.ids`.

The second important function is `create.fused`. This function needs again the recipient data set and the donor data set as input. Once again they are passed to the arguments `data.rec` and `data.don`, respectively. Further arguments that have to be specified are `mtc.ids` and `z.vars`. `mtc.ids` requires exactly that kind of a 2-column matrix as obtained in the output of `RANDwNND.hotdeck` (list element `mtc.ids` of the output). To the argument `z.vars` are passed the names of those variables where the missing values should be imputed, hence the names of variables available only in the donor data set. The output is the imputed recipient data set.

The rest of the coding regarding weighted random hot deck passes off analogously to the other implemented model: The at-risk-of-poverty rates for Austria and the nine states get computed as weighted means or with the help of the function `arpr`.

The R-Code regarding weighted random hot deck imputation of `ARPT60i` and `EQ_INC`, respectively, can be found in Listing 20, Listing 22 and Listing 21, Listing 23, respectively. Listing 20 and Listing 21 correspond to the versions where the variable state is used for the building of donation classes and Listing 22 and Listing 23 correspond to the versions with the usage of variable state and foreign origin of the household for the building of domains.

Listing 20: Weighted Random Hot Deck: imputation of `ARPT60i` in the Micro Census (**MZ** or **MZ**[*] equates to `samp[[1]]`) with the usage of EU-SILC (**S** or **S**[*] equates to `samp[[2]]`) and calculation of arpr out of the imputed values: with the usage of the variable state for building donation classes

```
match_smr_strat_arpt <- function(samp){
    names(samp[[1]])[which(names(samp[[1]])=="ARPT60i")] <- "
        ARPT60iorig"
    # exclude observations with weight 0 in both data sets
    samp[[1]] <- samp[[1]][samp[[1]]$gew!=0,]
    samp[[2]] <- samp[[2]][samp[[2]]$gew!=0,]

    group.v1 <- "strat"   # variable for building domains
    X.mtc <- NULL   # no matching variables
    # imputation of ARPT60i within the domains using weights:
    rnd.1 <- RANDwNND.hotdeck(data.rec=samp[[1]], data.don=samp
        [[2]], match.vars=X.mtc, don.class=group.v1, dist.fun="Gower
        ", cut.don="k.dist", k=1, weight.don="gew")
    fA.knnd1 <- create.fused(data.rec=samp[[1]], data.don=samp
        [[2]], mtc.ids=rnd.1$mtc.ids, z.vars="ARPT60i")

    # estimate the at-risk-of-poverty rate for Austria + 9 states
    ARPR <- weighted.mean(as.numeric(fA.knnd1$ARPT60i)-1, weights="
        gew")
    for (b in 1:9){
        bb <- levels(samp[[1]]$strat)[b]   # state name
        # limit the data set and weight vector with samp[[1]]$
            strat==bb
```

86

```
            ARPR <- c(ARPR, weighted.mean(as.numeric(fA.knnd1[
                fA.knnd1$strat==bb,]$ARPT60i)-1, weights="gew"))
    }
    ARPR
}
```

Listing 21: Weighted Random Hot Deck: imputation of EQ_INC in the Micro Census (MZ or MZ$^{(*)}$ equates to samp[[1]]) with the usage of EU-SILC (S or S$^{(*)}$ equates to samp[[2]]) and calculation of arpr out of the imputed values: with the usage of the variable state for building donation classes

```
match_smr_strat_inc <- function(samp){
    names(samp[[1]])[which(names(samp[[1]])=="EQ_INC")] <- "EQ_
        INCorig"
    # exclude observations with weight 0 in both data sets
    samp[[1]] <- samp[[1]][samp[[1]]$gew!=0,]
    samp[[2]] <- samp[[2]][samp[[2]]$gew!=0,]

    group.v1 <- "strat"    # variable for building domains
    X.mtc <- NULL   # no matching variables
    # imputation of EQ_INC within the domains using weights
    rnd.1 <- RANDwNND.hotdeck(data.rec=samp[[1]], data.don=samp
        [[2]], match.vars=X.mtc, don.class=group.v1, dist.fun="Gower
        ", cut.don="k.dist", k=1, weight.don="gew")
    fA.knnd1 <- create.fused(data.rec=samp[[1]], data.don=samp
        [[2]], mtc.ids=rnd.1$mtc.ids, z.vars="EQ_INC")

    # estimate the at-risk-of-poverty rate for Austria + 9 states
    ARPR <- c(arpr("EQ_INC", weights = "gew", breakdown = "strat",
        data = fA.knnd1)$value, arpr("EQ_INC", weights = "gew",
        breakdown = "strat", data = fA.knnd1)$valueByStratum$value)
        /100
    ARPR
}
```

Listing 22: Weighted Random Hot Deck: imputation of ARPT60i in the Micro Census (MZ or MZ$^{(*)}$ equates to samp[[1]]) with the usage of EU-SILC (S or S$^{(*)}$ equates to samp[[2]]) and calculation of arpr out of the imputed values: with the usage of the variables state and foreign origin of the household for building donation classes

```
match_smr_stratmig_arpr <- function(samp){
    names(samp[[1]])[which(names(samp[[1]])=="ARPT60i")] <- "
        ARPT60iorig"
    # exclude observations with weight 0 in both data sets
    samp[[1]] <- samp[[1]][samp[[1]]$gew!=0,]
    samp[[2]] <- samp[[2]][samp[[2]]$gew!=0,]

    group.v2 <- c("strat", "migration")    # variables for building
        domains
    X.mtc <- NULL   # no matching variables
```

```
# imputation of ARPT60i within the domains using weights:
rnd.2 <- RANDwNND.hotdeck(data.rec=samp[[1]], data.don=samp
    [[2]], match.vars=X.mtc, don.class=group.v2, dist.fun="Gower
    ", cut.don="k.dist", k=1, weight.don="gew")
fA.knnd2 <- create.fused(data.rec=samp[[1]], data.don=samp
    [[2]], mtc.ids=rnd.2$mtc.ids, z.vars="ARPT60i")

# estimate the at-risk-of-poverty rate for Austria + 9 states
ARPR <- weighted.mean(as.numeric(fA.knnd2$ARPT60i)-1, weights="
    gew")
for (b in 1:9){
        bb <- levels(samp[[1]]$strat)[b]    # state name
        # limit the data set and weight vector with samp[[1]]$
            strat==bb
        ARPR <- c(ARPR, weighted.mean(as.numeric(fA.knnd2[
            fA.knnd2$strat==bb,]$ARPT60i)-1, weights="gew"))
}
ARPR
}
```

Listing 23: Weighted Random Hot Deck: imputation of EQ_INC in the Micro Census (**MZ** or **MZ**[(*)] equates to samp[[1]]) with the usage of EU-SILC (**S** or **S**[(*)] equates to samp[[2]]) and calculation of arpr out of the imputed values: with the usage of the variables state and foreign origin of the household for building donation classes

```
match_smr_stratmig_inc <- function(samp){
    names(samp[[1]])[which(names(samp[[1]])=="EQ_INC")] <- "EQ_
        INCorig"
    # exclude observations with weight 0 in both data sets
    samp[[1]] <- samp[[1]][samp[[1]]$gew!=0,]
    samp[[2]] <- samp[[2]][samp[[2]]$gew!=0,]

    group.v2 <- c("strat", "migration")    # variables for building
        domains
    X.mtc <- NULL    # no matching variables
    # imputation of EQ_INC within the domains using weights
    rnd.2 <- RANDwNND.hotdeck(data.rec=samp[[1]], data.don=samp
        [[2]], match.vars=X.mtc, don.class=group.v2, dist.fun="Gower
        ", cut.don="k.dist", k=1, weight.don="gew")
    fA.knnd2 <- create.fused(data.rec=samp[[1]], data.don=samp
        [[2]], mtc.ids=rnd.2$mtc.ids, z.vars="EQ_INC")

    # estimate the at-risk-of-poverty rate for Austria + 9 states
    ARPR <- c(arpr("EQ_INC", weights = "gew", breakdown = "strat",
        data = fA.knnd2)$value, arpr("EQ_INC", weights = "gew",
        breakdown = "strat", data = fA.knnd2)$valueByStratum$value)
        /100
    ARPR
}
```

All the so far listed functions, beginning with `sampling` (see Listing 1) and ending with `match_smr_stratmig_inc` (see Listing 23) are saved in the file "`functions.R`". This file will be used further down in the rest of the R-Code.

A.1.3. Additional R-Code

Due to the computational complexity, the computations are done in parallel on 15 processors. Thus a file has been written with code for the computation of the at-risk-of-poverty rates for the different EU-SILC and Micro Census samples as well as for all the bootstrap samples. This code is saved in "`overall.R`" (see Listing 24) and the file is called later for the parallel computation, so it is applied to all 15 processors, see Listing 25.

In Listing 24 the needed packages and the data are loaded first. There is a outer loop, the looping concerning the repeated drawing of the artificial samples **S** and **MZ** and a inner loop, the looping concerning the bootstrapping of an EU-SILC and a Micro Census sample. The number of drawing samples (corresponding to the outer loop) is set to 21 and the number of bootstrap replicates is set to 125.

In the outer loop one sample pair (**MZ** and **S**) is drawn with the above listed function `sampling` (see Listing 1), the weights are calculated for both samples and the variable at-risk-of-poverty is calculated for **S** out of the variable `EQ_INC` (again with the function `arpr`). Then for **MZ** the at-risk-of-poverty rates for Austria and the nine states get estimated using all the above described methods and models (see inter alia Listings 2 - 23).

For the inner loop the number of repetitions of the observations in the several bootstrap samples are chosen with the help of the function `subbootweights` out of the package `survey` [see Lumley, 2012, 2004], more precisely the weights are returned from this function. So in the inner loop an adjustment of the weights is necessary for the samples and the at-risk-of-poverty rates get calculated and estimated afresh.

Then the variance and the mean of the at-risk-of-poverty rates from the bootstrap replicates are calculated once for every EU-SILC data set and once for every Micro Census data set for every estimation (obtained from the several methods and models).

Furthermore the computation time gets saved.

The output of `overall.R` contains then the names of the several used methods and models, the at-risk-of-poverty rates estimated on the basis of the different samples of **S**, the at-risk-of-poverty rates estimated on the basis of the estimated values of the different samples of **MZ**, the variances and the means of the results from the bootstrap samples of **S** as well as of **MZ** (for every model) and the elapsed time for the computation.

Listing 24: `overall.R`: the R-Code for every node: The output contains the names of the several used methods and models, the arpr estimated on the basis of the different samples of **S**, the arpr estimated on the basis of the estimated values of the different samples of **MZ**, the variances and the means of the results from the bootstrap samples of **S** as well as **MZ** (for every model) and the elapsed time for the computation.

```
library(laeken)
library(sampling)
library(survey)
library(robustbase)
library(MASS)
library(lme4)
library(VIM)
library(StatMatch)


load("pop_manip.RData")
load("smhv_2.RData")
smhv_2_2011 <- smhv_2[smhv_2$year==2011,]


REPSAMPLE <- 21     # number of drawing samples
REPBOOT <- 125      # number of bootstrap replicates

s <- arprS <- arprMZ <- varBS <- meanBS <- varBMZ <- meanBMZ <- vector(
    REPSAMPLE, mode = "list")
arprSb <- arprMZb <- vector(REPBOOT, mode = "list")

ModSch <- list(gewlmreg_soz=gewlmreg_soz, gewlmreg_rV=gewlmreg_rV,
    gewlmreg_sozrob=gewlmreg_sozrob, gewlmreg_rVrob=gewlmreg_rVrob,
    logreg_soz=logreg_soz, logreg_rV=logreg_rV, mixed_lm_inc_soz=mixed_
    lm_inc_soz, mixed_lm_inc_rV=mixed_lm_inc_rV, mixed_logm_arpt_soz=
    mixed_logm_arpt_soz, mixed_logm_arpt_rV=mixed_logm_arpt_rV, match_
    strat_arpt=match_strat_arpt, match_strat_inc=match_strat_inc, match_
    stratmigr_arpt=match_stratmigr_arpt, match_stratmigr_inc=match_
    stratmigr_inc, match_stratmigr_arpt_feklaw=match_stratmigr_arpt_
    feklaw, match_stratmigr_inc_feklaw=match_stratmigr_inc_feklaw, match
    _stratmigr_arpt_ALLE=match_stratmigr_arpt_ALLE, match_stratmigr_inc_
    ALLE=match_stratmigr_inc_ALLE, match_smr_strat_arpt=match_smr_strat_
    arpt, match_smr_strat_inc=match_smr_strat_inc, match_smr_stratmig_
    arpr=match_smr_stratmig_arpr, match_smr_stratmig_inc=match_smr_
    stratmig_inc)
modnames <- c("gewlmreg_soz", "gewlmreg_rV", "gewlmreg_sozrob", "
    gewlmreg_rVrob", "logreg_soz", "logreg_rV", "mixed_lm_inc_soz", "
    mixed_lm_inc_rV", "mixed_logm_arpt_soz", "mixed_logm_arpt_rV", "
    match_strat_arpt", "match_strat_inc", "match_stratmigr_arpt", "match
    _stratmigr_inc", "match_stratmigr_arpt_feklaw", "match_stratmigr_inc
    _feklaw", "match_stratmigr_arpt_ALLE", "match_stratmigr_inc_ALLE", "
    match_smr_strat_arpt", "match_smr_strat_inc", "match_smr_stratmig_
    arpr", "match_smr_stratmig_inc")


ptm <- proc.time()
for (i in 1:REPSAMPLE){
        s[[i]] <- samples(eusilcP_manip, smhv_2_2011)  # contains a
```

```
list of 2 (MZ equates s[[i]][[1]] und S equates s[[i]][[2]])
    for every i
# computation of the weights:
s[[i]][[1]]$gew <- 1/(s[[i]][[1]]$Prob)
s[[i]][[2]]$gew <- 1/(s[[i]][[2]]$Prob)
# computation of the thresholds for the computation of the
    variable at-risk-of-poverty:
ts1 <- arpr("EQ_INC", weights="gew", breakdown="strat", data=s
    [[i]][[1]])$threshold
ts2 <- arpr("EQ_INC", weights="gew", breakdown="strat", data=s
    [[i]][[2]])$threshold
# computation of the variable at-risk-of-poverty:
s[[i]][[1]]$ARPT60i <- as.factor(as.numeric(s[[i]][[1]]$EQ_INC
    < ts1))
s[[i]][[2]]$ARPT60i <- as.factor(as.numeric(s[[i]][[2]]$EQ_INC
    < ts2))

# estimation of arpr for S (for Austria and the 9 states):
arprS[[i]] <- c(weighted.mean(as.numeric(s[[i]][[2]][,"ARPT60i"
    ])-1,s[[i]][[2]]$gew), arpr("EQ_INC", weights="gew",
    breakdown = "strat", data = s[[i]][[2]])$valueByStratum$
    value /100 )
# estimation of arpr for MZ with every described method/model (
    for Austria and the 9 states):
arprMZ[[i]] <- lapply(ModSch, function(x) x(s[[i]]))

# number of repetitions of the observations in a bootstrap
    sample:
anz_in_bsamp_s <- data.frame(subbootweights(s[[i]][[2]]$strat,
    s[[i]][[2]]$ID_unit, replicates = REPBOOT, compress=FALSE)
    [[1]])
anz_in_bsamp_mz <- data.frame(subbootweights(s[[i]][[1]]$strat,
    s[[i]][[1]]$ID_unit, replicates = REPBOOT, compress=FALSE)
    [[1]])

for (j in 1:REPBOOT){
        sb <- s[[i]]
        # adaption of the weights for the different bootstrap
            runs
        sb[[1]]$gew <- s[[i]][[1]]$gew*anz_in_bsamp_mz[,j]
        sb[[2]]$gew <- s[[i]][[2]]$gew*anz_in_bsamp_s[,j]

        # new computation of the thresholds and the variable at
            -risk-of-poverty:
        ts1 <- arpr("EQ_INC", weights="gew", breakdown="strat",
            data=sb[[1]])$threshold
        ts2 <- arpr("EQ_INC", weights="gew", breakdown="strat",
            data=sb[[2]])$threshold
        sb[[1]]$ARPT60i <- as.factor(as.numeric(sb[[1]]$EQ_INC
            < ts1))
        sb[[2]]$ARPT60i <- as.factor(as.numeric(sb[[2]]$EQ_INC
            < ts2))

        # estimation of arpr for S and MZ (for every method/
```

```
                model) (for Austria and the 9 states):
        arprSb[[j]] <- c(weighted.mean(as.numeric(sb[[2]][,"
            ARPT60i"])-1,sb[[2]]$gew), arpr("EQ_INC", weights="
            gew", breakdown = "strat", data = sb[[2]])$
            valueByStratum$value /100)
        arprMZb[[j]] <- lapply(ModSch, function(x) x(sb))
    }

    varBS[[i]] <- apply(matrix(unlist(arprSb), nrow=10),1,var)
    meanBS[[i]] <- apply(matrix(unlist(arprSb), nrow=10),1,mean)
    varBMZ[[i]] <- vector(length(ModSch), mode="list")
    meanBMZ[[i]] <- vector(length(ModSch), mode="list")
    names(varBMZ[[i]]) <- modnames
    names(meanBMZ[[i]]) <- modnames
    for (k in 1:length(ModSch)){
        varBMZ[[i]][[k]] <- apply(matrix(unlist(lapply(arprMZb,
            function(x) x[[k]])), nrow=10),1,var)
        meanBMZ[[i]][[k]] <- apply(matrix(unlist(lapply(arprMZb
            , function(x) x[[k]])), nrow=10),1,mean)
    }
}
verstrichen <- (proc.time() - ptm)["elapsed"]

par_erg <- list(modnames=modnames, arprS=arprS, arprMZ=arprMZ, varBS=
    varBS, varBMZ=varBMZ, meanBS=meanBS, meanBMZ=meanBMZ, verstrichen=
    verstrichen)
par_erg
```

The parallelization (see Listing 25) is done with functions from the packages snow [see Tierney et al., 2013] and rlecuyer [see Sevcikova and Rossini, 2012].

With the function makeCluster (out of the package snow) a "SNOW Cluster" is started and in the end the cluster gets shut down with the function stopCluster (out of the same package snow). The input of the function makeCluster is among others the number of slaves nodes and this number is set to 15 here. The used cluster type is SOCK, see Listing 25.

The function clusterEvalQ out of the package snow is a function for computing on a SNOW cluster and it becomes as input a cluster object and an expression to evaluate [see Tierney et al., 2013]. The used expression is source('' ... .R''), so the R-File '' ... .R'' is loaded and executed on each cluster node.

The function clusterSetupRNG loads the package rlecuyer and handles the uniform random number generation in SNOW Clusters.

The function system.time is used to determine the elapsed time.

Then the results obtained for all nodes are combined and quality criteria as the variance and the bias, the MSE, a bias corrected version of the bias and the MSE (by using the mean of the bootstrap replicates for the computation of the bias), the difference in the mean of the variances yield from the bootstrap replicates and the variance calculated on the basis of the results of the repeated sample drawing and the difference in the mean of the at-risk-of-poverty rates yield from the bootstrap replicates (of all drawn samples) and the mean of the at-risk-of-poverty rates calculated on the basis of the results of the repeated sample drawing are computed (see Listing 25).

Listing 25: Parallelization of the R-Code of `overall.R` and computing of the different quality criteria.

```
library(snow)
CLREP <- 15      # number of nodes
cl <- makeCluster(CLREP, type = "SOCK") # start SNOW cluster
# load all the functions
clusterEvalQ(cl, source("/home/verena/functions.R"))
set.seed(826296)
library(rlecuyer)
clusterSetupRNG(cl, seed=12345)
# load and execute the R-Code of overall.R:
system.time(g <- clusterEvalQ(cl, source("/home/verena/overall.R")))
stopCluster(cl) # stop SNOW cluster

modnames <- g[[1]][[1]][[1]]
# combination of the results obtained from the several nodes:
arprS <- arprMZ <- varBS <- varBMZ <- meanBS <- meanBMZ <- verstrichen
    <- NULL
for (i in 1:CLREP){
        arprS <- c(arprS, g[[i]][[1]][[2]])
        arprMZ <- c(arprMZ, g[[i]][[1]][[3]])
        varBS <- c(varBS, g[[i]][[1]][[4]])
        varBMZ <- c(varBMZ, g[[i]][[1]][[5]])
        meanBS <- c(meanBS, g[[i]][[1]][[6]])
        meanBMZ <- c(meanBMZ, g[[i]][[1]][[7]])
        verstrichen <- c(verstrichen, g[[i]][[1]][[8]])
}

# calculation of the standard deviation/variance of the arpr
    estimations
varS <- apply(matrix(unlist(arprS), nrow=10), 1, var)
SDS <- sqrt(varS)
varMZ <- SDMZ <- vector(length(modnames), mode="list")
names(varMZ) <- names(SDMZ) <- modnames
for (k in 1:length(modnames)){
        varMZ[[k]] <- apply(matrix(unlist(lapply(arprMZ, function(x) x
            [[k]])), nrow=10), 1, var)
        SDMZ[[k]] <- apply(matrix(unlist(lapply(arprMZ, function(x) x[[
            k]])), nrow=10), 1, sd)
}

# calculation of the bias of the arpr estimations
library(laeken)
load("pop_manip.RData")
arprPOP <- c(arpr("EQ_INC", breakdown="strat", data=eusilcP_manip)$
    value, arpr("EQ_INC", breakdown="strat", data=eusilcP_manip)$
    valueByStratum$value) /100
BiasS <- apply(matrix(unlist(arprS), nrow=10), 1, mean) - arprPOP
```

```
BiasMZ <- vector(length(modnames), mode="list")
names(BiasMZ) <- modnames
for (k in 1:length(modnames)){
        BiasMZ[[k]] <- apply(matrix(unlist(lapply(arprMZ, function(x) x
        [[k]])),nrow=10), 1, mean) - arprPOP }

# calculation of the MSE
MSES <- BiasS^2 + varS
MSEMZ <- vector(length(modnames), mode="list")
names(MSEMZ) <- modnames
for (k in 1:length(modnames)){
        MSEMZ[[k]] <- BiasMZ[[k]]^2 + varMZ[[k]]
}

# proofing the reliability of bootstrapping
DiffSDBSquerSDS <- apply(matrix(sqrt(unlist(varBS)), nrow=10), 1, mean)
        - SDS
DiffSDBMZquerSDMZ <- vector(length(modnames), mode="list")
names(DiffSDBMZquerSDMZ) <- modnames
for (k in 1:length(modnames)){
        DiffSDBMZquerSDMZ[[k]] <- apply(matrix(sqrt(unlist(lapply(
        varBMZ, function(x) x[[k]]))), nrow=10), 1, mean) - SDMZ[[k
        ]]
}

DiffMeanBSquerArprSquer <- apply(matrix(unlist(meanBS), nrow=10), 1,
        mean) - apply(matrix(unlist(arprS), nrow=10), 1, mean)
DiffMeanBMZquerArprMZquer <- vector(length(modnames), mode="list")
names(DiffMeanBMZquerArprMZquer) <- modnames
for (k in 1:length(modnames)){
        DiffMeanBMZquerArprMZquer[[k]] <- apply(matrix(unlist(lapply(
        meanBMZ, function(x) x[[k]])), nrow=10), 1, mean) - apply(
        matrix(unlist(lapply(arprMZ, function(x) x[[k]])), nrow=10),
        1, mean)
}

# calculation of a "corrected" version of the bias
BiasBS <- apply(matrix(unlist(meanBS), nrow=10), 1, mean) - arprPOP
BiasBMZ <- vector(length(modnames), mode="list")
names(BiasBMZ) <- modnames
for (k in 1:length(modnames)){
        BiasBMZ[[k]] <- apply(matrix(unlist(lapply(meanBMZ, function(x)
        x[[k]])), nrow=10), 1, mean) - arprPOP
}

# calculation of a bias corrected version of the MSE
MSEBS <- BiasBS^2 + varS
MSEBMZ <- vector(length(modnames), mode="list")
names(MSEBMZ) <- modnames
for (k in 1:length(modnames)){
        MSEBMZ[[k]] <- BiasBMZ[[k]]^2 + varMZ[[k]]
}
```

A.2. Additional Tables

In this chapter the tables refered to in Chapter 4.3 concerning the averages (of the nine states or of the several models) for the several quality criteria are listed. For comments and reading it is refered to Chapter 4.3.

(a)(i) Linear Regression	0.10153	(f)(i) Random Hot Deck	0.21669
(a)(ii) Linear Regression	0.07086	(f)(ii) Random Hot Deck	0.20985
(b)(i) Robust Lin.Regr.	0.00830	(f)(iii) Random Hot Deck	0.17721
(b)(ii) Robust Lin.Regr.	-0.06732	(f)(iv) Random Hot Deck	0.19706
(c)(i) Logistic Regr.	0.51222	(g)(i) Sequential R.H.D.	-0.28714
(c)(ii) Logistic Regr.	0.36016	(g)(ii) Sequential R.H.D.	-0.25378
(d)(i) Linear Mixed Regr.	0.10532	(g)(iii) Sequential R.H.D.	-0.04083
(d)(ii) Linear Mixed Regr.	0.07086	(g)(iv) Sequential R.H.D.	-0.01558
(e)(i) Logistic Mixed Regr.	0.47969	(h)(i) Weighted R.H.D.	0.13202
(e)(ii) Logistic Mixed Regr.	0.39836	(h)(ii) Weighted R.H.D.	0.18083
		(h)(iii) Weighted R.H.D.	0.15305
		(h)(iv) Weighted R.H.D.	0.20765

Table 30: Average (of the nine states) of the difference in the mean of the standard deviations yield from the bootstrap replicates and the standard deviation calculated on the basis of the results of the repeated sample drawing (see also Equation 8 and Equation 9) for the several models [in %].

Austria	0.01324
Burgenland	0.08966
Lower Austria	-0.00136
Vienna	-0.01986
Carinthia	0.03576
Styria	0.01511
Upper Austria	-0.02082
Salzburg	0.03632
Tyrol	0.05698
Vorarlberg	0.06006

Table 31: Mean (of the several models) of the difference in the mean of the at-risk-of-poverty rates yield from the bootstrap replicates (of all drawn samples) and the mean of the at-risk-of-poverty rates estimated on the basis of the results of the repeated sample drawing (see also Equation 10 and Equation 11) for the nine states and Austria [in %].

method	Av.bias	Av.abs.bias	Av.corr.bias	Av.abs.corr.bias
(a)(i) Linear Regression	-3.83234	3.88471	-3.55436	3.65645
(a)(ii) Linear Regression	-3.96156	3.96156	-3.62650	3.62650
(b)(i) Robust Lin. Regr.	-5.78892	5.78892	-6.32526	6.32526
(b)(ii) Robust Lin. Regr.	-5.92280	5.92280	-6.42301	6.42301
(c)(i) Logistic Regr.	-0.94639	2.16821	-0.88297	2.17633
(c)(ii) Logistic Regr.	0.12549	0.13114	0.18940	0.18940
(d)(i) Linear Mixed Regr.	-3.76573	3.82880	-3.49547	3.60934
(d)(ii) Linear Mixed Regr.	-3.96156	3.96156	-3.62650	3.62650
(e)(i) Logistic Mixed Regr.	-0.17001	2.17527	-0.16121	2.20388
(e)(ii) Logistic Mixed Regr.	0.11963	0.11963	0.18897	0.18897
(f)(i) Random Hot Deck	0.10627	0.11596	0.18323	0.18323
(f)(ii) Random Hot Deck	0.15919	0.18451	0.25399	0.25399
(f)(iii) Random Hot Deck	0.09498	0.13730	0.19774	0.19774
(f)(iv) Random Hot Deck	0.17043	0.18576	0.24474	0.24474
(g)(i) Sequential R. H. D.	1.42645	1.42645	1.15602	1.15602
(g)(ii) Sequential R. H. D.	0.60647	0.61851	0.43973	0.45682
(g)(iii) Sequential R. H. D.	0.05950	0.17791	0.08248	0.17125
(g)(iv) Sequential R. H. D.	0.17387	0.20218	0.21318	0.21981
(h)(i) Weighted R. H. D.	0.12382	0.12506	0.19198	0.19198
(h)(ii) Weighted R. H. D.	0.17550	0.17550	0.21844	0.21844
(h)(iii) Weighted R. H. D.	0.16291	0.16291	0.18941	0.18941
(h)(iv) Weighted R. H. D.	0.12898	0.12898	0.21251	0.21251

Table 32: In the second column the average (of the nine states) of the bias for every model [in %] and in the third column the average (of the nine states) of the absolute values of the bias for every model [in %] are listed. In the fourth column the average (of the nine states) of the corrected bias for every model [in %] and in the fifth column the average (of the nine states) of the absolute values of the corrected bias for every model [in %] are listed.

direct estimator (EU-SILC) SD$_S$	1.49464	(f)(i) Random Hot Deck	2.22797
(a)(i) Linear Regression	0.93481	(f)(ii) Random Hot Deck	2.30081
(a)(ii) Linear Regression	1.39931	(f)(iii) Random Hot Deck	2.23672
(b)(i) Robust Lin. Regr.	0.72920	(f)(iv) Random Hot Deck	2.27670
(b)(ii) Robust Lin. Regr.	1.00472	(g)(i) Sequential R. H. D.	2.53439
(c)(i) Logistic Regr.	0.88192	(g)(ii) Sequential R. H. D.	2.50240
(c)(ii) Logistic Regr.	1.62615	(g)(iii) Sequential R. H. D.	2.06541
(d)(i) Linear Mixed Regr.	0.93412	(g)(iv) Sequential R. H. D.	2.11465
(d)(ii) Linear Mixed Regr.	1.39931	(h)(i) Weighted R. H. D.	1.67621
(e)(i) Logistic Mixed Regr.	0.97009	(h)(ii) Weighted R. H. D.	1.70655
(e)(ii) Logistic Mixed Regr.	1.60463	(h)(iii) Weighted R. H. D.	1.64022
		(h)(iv) Weighted R. H. D.	1.66054

Table 33: Average (of the nine states) of the standard deviation for the direct estimator and every model [in %].

method	av.MSE	av.corr.MSE
direct estimator (EU-SILC)	2.597e-04	2.619e-04
(a)(i) Linear Regression	0.00253	0.00231
(a)(ii) Linear Regression	0.00276	0.00246
(b)(i) Robust Lin. Regr.	0.00432	0.00494
(b)(ii) Robust Lin. Regr.	0.00459	0.00515
(c)(i) Logistic Regr.	9.263e-04	9.066e-04
(c)(ii) Logistic Regr.	3.030e-04	3.047e-04
(d)(i) Linear Mixed Regr.	0.00251	0.00229
(d)(ii) Linear Mixed Regr.	0.00276	0.00246
(e)(i) Logistic Mixed Regr.	8.498e-04	8.665e-04
(e)(ii) Logistic Mixed Regr.	2.930e-04	2.957e-04
(f)(i) Random Hot Deck	5.590e-04	5.610e-04
(f)(ii) Random Hot Deck	6.115e-04	6.136e-04
(f)(iii) Random Hot Deck	5.715e-04	5.735e-04
(f)(iv) Random Hot Deck	5.852e-04	5.877e-04
(g)(i) Sequential R. H. D.	9.025e-04	8.233e-04
(g)(ii) Sequential R. H. D.	7.264e-04	6.968e-04
(g)(iii) Sequential R. H. D.	4.806e-04	4.805e-04
(g)(iv) Sequential R. H. D.	5.078e-04	5.094e-04
(h)(i) Weighted R. H. D.	3.147e-04	3.172e-04
(h)(ii) Weighted R. H. D.	3.321e-04	3.336e-04
(h)(iii) Weighted R. H. D.	3.033e-04	3.044e-04
(h)(iv) Weighted R. H. D.	3.101e-04	3.130e-04

Table 34: In the second column the average (of the nine states) of the mean squared error for the direct estimator and every model and in the third column the average (of the nine states) of the bias corrected mean squared error for the direct estimator and every model are listed.

97

References

A. Alfons and S. Kraft. *simPopulation: Simulation of synthetic populations for surveys based on sample data*, 2012. URL http://CRAN.R-project.org/package=simPopulation. R package version 0.4.0.

A. Alfons and M. Templ. Estimation of social exclusion indicators from complex surveys: The r package laeken. *Journal of Statistical Software*, 54(15), 2013. URL http://www.jstatsoft.org/v54/i15/.

A. Alfons, S. Kraft, M. Templ, and P. Filzmoser. Simulation of close-to-reality population data for household surveys with application to eu-silc. *Statistical Methods and Applications*, 20(3):383–407, 2011.

A. Alfons, J. Holzer, and M. Templ. *laeken: Estimation of indicators on social exclusion and poverty*, 2013. URL http://CRAN.R-project.org/package=laeken. R package version 0.4.4.

D. Bates, M. Maechler, and B. Bolker. *lme4: Linear mixed-effects models using S4 classes*, 2013. URL http://CRAN.R-project.org/package=lme4. R package version 0.999999-2.

M. Bauer, M. Till, R. Heuberger, M. Bilgili, T. Glaser, E. Kafka, J. Klotz, A. Kowarik, N. Lamei, A. Meraner, A. Oismüller, M. Plate, S. Scheikl, and V. Zucha. Studie zu Armut und sozialer Eingliederung in den Bundesländern. Technical report, Statistics Austria, 2013.

R. Christensen. *Plane Answers to Complex Questions: The Theory of Linear Models*. Springer Texts in Statistics. Springer New York, 2011. ISBN 9781441998163.

M. D'Orazio. *StatMatch: Statistical Matching*, 2012. URL http://CRAN.R-project.org/package=StatMatch. R package version 1.2.0.

M. D'Orazio, M. Di Zio, and M. Scanu. *Statistical Matching: Theory and Practice*. Wiley Series in Survey Methodology. John Wiley & Sons, 2006. ISBN 9780470023532.

B. Efron and R. Tibshirani. *An Introduction to the Bootstrap*. Monographs on statistics and applied probabilities. Chapman & Hall/CRC, 1993. ISBN 9780412042317.

T. Glaser and R. Heuberger. Standard-Dokumentation Metainformationen (Definitionen, Erläuterungen, Methoden, Qualität) zu EU-SILC 2010. Technical report, Statistics Austria, Vienna, Austria, 2012.

T.J. Hastie, R.J. Tibshirani, and J.J.H. Friedman. *The Elements of Statistical Learning: Data Mining, Inference, and Prediction*. Springer series in statistics. Springer-Verlag New York, 2009. ISBN 9780387848587. URL http://www-stat.stanford.edu/ElemStatLearn.

S.G. Heeringa, B.T. West, and P.A. Berglund. *Applied Survey Data Analysis*. Chapman & Hall/CRC Statistics in the Social and Behavioral Sciences Series. Taylor & Francis Group, 2010. ISBN 9781420080667.

R.A. Johnson and D.W. Wichern. *Applied Multivariate Statistical Analysis*. Prentice-Hall, 4th edition, 1998. ISBN 9780138341947. ISBN 0-13-834194-X.

R. Lehtonen, A. Veijanen, M. Myrskylä, and M. Valaste. Small area estimation of indicators on poverty and social exclusion. Research Project Report WP2 – D2.2, FP7-SSH-2007-217322 AMELI, 2011. URL http://ameli.surveystatistics.net.

T. Lumley. Analysis of complex survey samples. *Journal of Statistical Software*, 9(1): 1–19, 2004. R package verson 2.2.

T. Lumley. survey: analysis of complex survey samples, 2012. R package version 3.28-2.

W.G. Madow and I. Olkin. *Incomplete Data in Sample Surveys: Proceedings of the Symposium*, volume 3. Academic Press, 1983. ISBN 0123639034.

W.G. Madow, I. Olkin, and D.B. Rubin. *Incomplete Data in Sample Surveys: Theory and Bibliographies*, volume 2. Academic Press, 1983. ISBN 0123639026.

C. Moser, M. Fasching, V. Zucha, and Klapfer K. Standard-Dokumentation Metainformationen (Definitionen, Erläuterungen, Methoden, Qualität) zu Mikrozensus ab 2004 - Arbeitskräfte- und Wohnungserhebung. Technical report, Statistics Austria, Vienna, Austria, 2013.

R Core Team. *R: A Language and Environment for Statistical Computing*. R Foundation for Statistical Computing, Vienna, Austria, 2013. URL http://www.R-project.org/.

H. Rinne. *Taschenbuch der Statistik*, volume 3. Wissenschaftlicher Verlag Harri Deutsch Gmbh, 2003. ISBN 3817116950.

P. Rousseeuw, C. Croux, V Todorov, A. Ruckstuhl, M. Salibian-Barrera, T. Verbeke, M. Koller, and M. Maechler. *robustbase: Basic Robust Statistics*, 2013. URL http://CRAN.R-project.org/package=robustbase. R package version 0.9-8.

P.J. Rousseeuw and A.M. Leroy. *Robust Regression and Outlier Detection*. Wiley Series in Probability and Statistics. John Wiley & Sons, 2003. ISBN 9780471488552.

L. Sachs and J. Hedderich. *Angewandte Statistik: Methodensammlung mit R*. Springer, 13 edition, 2009. ISBN 9783540889014.

H. Sevcikova and T. Rossini. *rlecuyer: R interface to RNG with multiple streams*, 2012. URL http://CRAN.R-project.org/package=rlecuyer. R package version 0.3-3.

M. Templ and A. Alfons. Variance estimation of indicators on social exclusion and poverty using the r package laeken. Technical report, 2013. URL http://CRAN.R-project.org/package=laeken. R package version 0.4.4.

M. Templ, A. Alfons, A. Kowarik, and B. Prantner. *VIM: Visualization and Imputation of Missing Values*, 2013. URL http://CRAN.R-project.org/package=VIM. R package version 3.0.3.1.

T. Tierney, A.J. Rossini, N. Li, and H. Sevcikova. *snow: Simple Network of Workstations*, 2013. URL http://CRAN.R-project.org/package=snow. R package version 0.3-12.

Y. Tillé and A. Matei. *sampling: Survey Sampling*, 2012. URL http://CRAN.R-project.org/package=sampling. R package version 2.5.

V. Todorov and P. Filzmoser. An object-oriented framework for robust multivariate analysis. *Journal of Statistical Software*, 32(3):1–47, 2009. URL http://www.jstatsoft.org/v32/i03/.

W.N. Venables and B.D. Ripley. *Modern Applied Statistics with S*. Springer, New York, fourth edition, 2002. URL http://www.stats.ox.ac.uk/pub/MASS4. ISBN 0-387-95457-0.

K.M. Wolter. *Introduction to Variance Estimation*. Statistics for Social and Behavioral Sciences. Springer, 2010. ISBN 9781441921970.